Particle Swarm Optimization: A Physics-Based Approach

Particle Swarm Optimizaton: A Physics-Based Approach

Said M. Mikki and Ahmed A. Kishk

ISBN: 978-3-031-00576-3   paperback

ISBN: 978-3-031-01704-9   ebook

DOI 10.1007/978-3-031-01704-9

A Publication in the Springer series

SYNTHESIS LECTURES ON COMPUTATIONAL ELECTROMAGNETICS

Lecture #20

Series Editor: Constantine A. Balanis, Arizona State University

Series ISSN

Synthesis Lectures on Computational Electromagnetics

Print 1932-1252   Electronic 1932-1716

Particle Swarm Optimization:
A Physics-Based Approach

Said M. Mikki and Ahmed A. Kishk
University of Mississippi

SYNTHESIS LECTURES ON COMPUTATIONAL ELECTROMAGNETICS #20

ABSTRACT This work aims to provide new introduction to the particle swarm optimization methods using a formal analogy with physical systems. By postulating that the swarm motion behaves similar to both classical and quantum particles, we establish a direct connection between what are usually assumed to be separate fields of study, optimization and physics. Within this framework, it becomes quite natural to derive the recently introduced quantum PSO algorithm from the Hamiltonian or the Lagrangian of the dynamical system. The physical theory of the PSO is used to suggest some improvements in the algorithm itself, like temperature acceleration techniques and the periodic boundary condition. At the end, we provide a panorama of applications demonstrating the power of the PSO, classical and quantum, in handling difficult engineering problems. The goal of this work is to provide a general multi-disciplinary view on various topics in physics, mathematics, and engineering by illustrating their interdependence within the unified framework of the swarm dynamics.

KEYWORDS

particle swarm optimization, swarm dynamics, computational electromagnetics, evolutionary computing, artificial intelligence, optimization algorithm

Contents

Preface

This book aims to introduce the particle swarm optimization method to the widest range possible of scientists and engineers, including those working in theoretical and computational areas who need efficient global research strategies to study difficult problems and to design physical devices. The core of the book is based on results obtained by the authors in their research conducted during the last three years, and includes some problems in the area of computational electromagnetics. The theme most likely to be encountered in the various topics addressed here is the strong analogy between optimization and physical systems, which we have tried to exploit both formally and informally in order to attain new insights on the existing optimization code and to devise new modifications.

Said M. Mikki and Ahmed A. Kishk

CHAPTER 1

Introduction

1.1 WHAT IS OPTIMIZATION?

1.1.1 General Consideration

Roughly speaking, we can consider optimization as a methodology that requires two fundamental elements: *adaptation* and *purpose*. Strictly speaking, we can define optimization as: (1) a systematic change, modification, adaptation of a process that aims to (2) achieve a pre-specified purpose. This purpose can be the maximum/minimum of a numerical function defined by the user. In other words, optimization must involve a teleological element of knowing *what* we are going to do (maximize a function) but an ignorance of the *how*, the general route leading to the goal. A practical resolution of this paradoxical situation is the essence of the entire area of research called optimization theory.

Based on this understanding, an *optimization algorithm* can be defined as a set of clear instructions specifying how to proceed, starting from certain initial conditions, until we reach the final goal imposed before. An important point to add here is that most of the time the algorithm fails to achieve the exact pre-specified goal, arriving instead to an approximation of the goal. In science and engineering applications, there is always a degree of tolerance allowing us to accept certain margins of errors. This means that an optimization goal is always ideal, or a purely mathematical condition, which can be realized in nature only through a certain model that tries to mimic the physical or mathematical process under consideration.

1.1.2 A Very Simple Optimization Problem

To demonstrate these basic ideas, let us start with a very simple example of optimization. Consider the following problem: Find the solution of the equation $x^2 = 2$. Knowing that the theory of real numbers permits the existence of two exact solutions $\pm\sqrt{2}$, we turn our attention to the problem of *how* to find good approximations for this solution. One can formulate the task in the following way: What is the best rational number x that will minimize the function $F(x) = |x^2 - 2|$? The function F is what we call the goal, the objective function, the optimization measure, or crudely, the purpose.

We can immediately infer from the discussion above that optimization is a *search* problem. This should always be put in our minds while reading or interpreting algorithms. Although the appearance of rigid mathematical statements gives the deceptive impression that everything is evolving in a precise and deterministic way, a true optimization method is nothing but an organized style of executing a search for something missing. Thus, it is not surprising to find that the first and most direct optimization method is called *random* search, which is basically moving in arbitrary steps within the solution space until you hit the right answers!

Now, clearly the range of the variable x is huge, encompassing the entire set of rational numbers (notice that in reality we approximate any irrational numbers by a rational number with finite number of digits). So the question that arises naturally now is how to restrict the range of x to reduce the effort of our search for the solution of the optimization problem. It turns out that there is no unique answer. Each problem poses its own restrictions on the allowable physical range of the parameters involved. Such range is called in optimization theory the *parameters space*. We define also the *fitness space* as the set of numbers specifying how successful is a set of parameters in the optimization problem. Using these two simple definitions, one can then visualize the optimization process as a search in the parameters space that aims to reach the maximum/minimum of the fitness space [1]. In other words, the concept of fitness space can be employed to convert all symbolic optimization problems into numerical optimization problems.

Going back to our simple example, we see that the fitness measure is the function $F(x)$ while the optimization parameter is x. The range of x is—theoretically—the entire set of rational numbers, but for practical purposes we need to choose a smaller range. Since intuitively we do not expect the rational number approximating $\sqrt{2}$ to be far away from 2 itself, we pick the interval [0, 3], where we are interested here only in the positive root. Figure 1.1 shows the graph of the function $F(x) = |x^2 - 2|$. It is clear that the minimum is located at our—theoretical—solution $x = \sqrt{2}$. But how can we reach there? Random Search is the simplest optimization algorithm that can be employed to find the minimum. We devise the following algorithm.

1. Start from one of the interval ends, say at $x_0 = 0$.

2. Update the new value of x by adding small increment Δx

$$x_n = x_{n-1} + \Delta x . \tag{1.1}$$

3. Evaluate and store the fitness function $F(x)$ at all $x = x_n, n = 1, 2, .., N$, until you cover the parameter space of x.

4. The desired solution will be $x_{optm} = x_m$ such that

$$F(x_m) = \min\{F(x_n)\}_{n=1}^{n=N} . \tag{1.2}$$

This simple algorithm can be refined if the number of steps is increased, that is, by decreasing Δx, with the obvious drawback of increasing the computational time.

The first thing we notice about this algorithm is its extreme simplicity. Actually, nothing is needed to be known *a priori* about the problem except the parameter space. As we will show throughout this book, this is a very attractive feature shared by the PSO algorithm. However, the main disadvantage of any random search algorithm is its poor computational performance. In particular, one should cover all interesting locations in the parameter space in order to be sure that a *global* optimum was obtained. Such a thing is certainly prohibited in practical problems. What we

---

[1]To be more accurate, one needs also to introduce another space called the *function space*, which is defined as the results of all functions defined in the optimization problem to produce the goal or the measure. However, for most practical situations, the function space and the fitness space are the same [3].

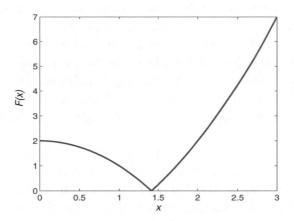

Figure 1.1: The graph of the function $F(x) = |x^2 - 2|$ with the minimum located at $x = \sqrt{2}$.

need, instead, is an algorithm "intelligent" enough to *know* where to head on in the fitness landscape. This is the deep connection between artificial intelligence research and optimization theory.

In the next chapters, we will introduce the PSO algorithm as a successful way of understanding the critical relation between the emergence of intelligent behavior in swarms of biological individuals and optimization in the sense described in this section. As is expected, it will be shown that this added intelligence will manifest itself in two modifications of the random search algorithm shown above. First, the increment Δx will be chosen "wisely" by adapting the swarm dynamics behavior to the optimization requirement as dictated by the objective function of the problem. Second, there will be no need to cover all of the points in the parameter space. The algorithm, if prepared properly, should be able to converge to the optimum solution. In other words, the increment Δx will go to zero when convergence is achieved so—in principle—one should not keep changing x_n until reaching the other end of the interval in which x is defined. Once the increment Δx is observed to be stabilizing at zero, the user can halt the algorithm and examine the results.

1.1.3   Exploration and Exploitation: The Fundamental Tradeoff

Depending on the landscape of the optimization problem, there are two types of optima: *global* and *local*. Local optima are maxima or minima only in a neighborhood of the point under consideration. Global optima are the maxima (or minima) of all the corresponding local maxima (or minima). This makes the global optima harder to obtain since one should—in principle—examine all the local structures of the landscape and then compare them with each other. This comparison process is in general very expensive and should be avoided. What we need is a global optimization search strategy that can converge to the global optima by considerations of the local landscape and without necessarily traversing all points in this landscape.

To understand the importance of this discussion, we introduce in this section two general concepts vital for the understanding of any optimization algorithm.

The first is *exploration*. This refers to the overall search for the approximate location where we expect the global optimum to be located in the fitness landscape. We say "approximate" because it turns out that a given algorithm is not capable of heading directly toward the global optima once the whereabouts of this optima have been determined. One should switch to another method that can refine the search, which is now already close enough to the right location, until good convergence is obtained.

The second process is what we call *exploitation*. It involves an iterative method starting from initial guess. The most basic of such methods is Hill Climbing Algorithms. There we update the position only if success (better fitness) is recorded. In this case, once the algorithm was started around certain optimum, convergence is guaranteed but not necessarily to the global optimum.

The main issue pertinent to us here is that there exists a fundamental tradeoff between exploration and exploitation. It is not possible to find a single algorithm capable of doing both at the same time. The PSO algorithm in this book is a strategy convenient for the global part of the optimization (exploration) but not always the best choice to achieve a refined estimation of the global optimum once its approximate location has been determined. Other methods may be used in *hybrid* combination. For example, the genetic algorithm (GA) or the PSO method can be combined with a local optimization code, like the gradient algorithm, to achieve the best possible performance.

1.2   WHY PHYSICS-BASED APPROACH

There is a close connection between the evolution of the dynamical variables in physical systems and optimization. It has been known since 200 years that the law of motion in Newtonian mechanics can be obtained by minimizing certain functionals called the "action." Moreover, with the invention of the path integral formalism by Richard Feynman in the last century, we now know that quantum phenomena can also be described by the very same approach [1]. It turns out that most of the known physical processes can be intimately related to some sort of "critical" or deeper variational problem: the optimum of this problem leads to the equations of motion in an elegant way.

Although this has been known for a long time, optimization is still considered in physics, applied mathematics, and engineering as a *tool* used to solve some difficult practical or theoretical problems. For example, the use of optimization codes in engineering problems is mainly due to our inability to find a desired solution in reasonable time. Optimization methods provide then a faster "search" for the best performance of the system of interest. However, optimization and physical systems are really two different ways to describe the same thing. If any particle moves in the trajectory that minimizes the action of the problem, then the search for the optimum of the functional (objective function) is equivalent to finding the equations of motion of the particle. Our point of view is that if a more formal analogy between physical systems and optimization algorithms can be constructed on the fundamental level, then deeper insights on both theoretical physics and the algorithm itself can be obtained.

1.3   THE PHILOSOPHY OF THE BOOK

The main purpose of this book is twofold. First, we take the particle swarm optimization (PSO) algorithm as a case study, or a toy model, to study how a Lagrangian formulation of this global optimization strategy can be constructed. This will motivate some of the tuning parameters that were introduced by the heuristic optimization community based on an intuition that is not directly related to the physical nature of the problem. Also, the Lagrangian formalism will reveal new hidden aspects in the algorithm. For example, we will show that the basic PSO algorithm lacks any "electromagnetic" nature. Particles are not electrically charged and hence do not radiate. However, the general formalism can provide some insights on how to add this electromagnetic behavior to the algorithm in future studies. The most interesting insight, however, is the possibility of looking to both the classical and quantum versions of the PSO algorithm as different manifestations of a single underlying Markov process, a view that is currently revived in theoretical physics.

Second, this work aims to introduce new inter-disciplinary perspectives for the artificial intelligence (AI) and evolutionary computing (EC) communities. While EC methods, generally referred to as *heuristic,* work very well with complicated problems, still little is known about the fundamental mechanisms responsible for the satisfactory performance of the technique. Although the social intelligence point of view has been emphasized considerably in the PSO literature, it seems that the treatment so far is still conventional in the sense of working with interpretations based on the previously established literature of the GA [3]. The main thrust behind EC and AI strategies is the inspiration by nature: in the case of AI nature is the human mind, while in EC it is the evolutionary paradigm. Therefore, it looks plausible to continue following this original impulse by allowing for further analogies inspired by similarities with other nonbiological natural phenomena. It is the hope of the authors of this work that the utilization of concepts that are totally outside the traditional AI literature may provide new routes in studying the problem of foundations of EC methods.

In the next chapters, we treat the PSO algorithm based on physics, rather than artificial intelligence (AI) or evolutionary computing (EC) [68]. The purpose of this analogy is to further enhance the understanding of how the algorithm works and to provide new tools for the study of the dynamics of the method. The basic idea is to observe the close similarity between a swarm of particles, communicating with each other through individual and social knowledge, and a collection of material particles interacting through a classic Newtonian field. In particular, molecular dynamics (MD) will be studied in connection to the PSO environment. Various relations and analogies will be established and illustrated throughout this book.

CHAPTER 2

The Classical Particle Swarm Optimization Method

2.1 DEFINITION OF THE PSO ALGORITHM

We start with an N-dimensional vector space $\mathbf{R}^N$. A population of M particles is assumed to evolve in this space such that each particle is assigned the following position and velocity vectors, respectively:

$$\mathbf{r}^i(t) = \left[\begin{array}{cccc} r_1^i(t) & r_2^i(t) & . \quad . \quad . & r_N^i(t) \end{array}\right]^T, \qquad (2.1)$$

$$\mathbf{v}^i(t) = \left[\begin{array}{cccc} v_1^i(t) & v_2^i(t) & . \quad . \quad . & v_N^i(t) \end{array}\right]^T, \qquad (2.2)$$

where T is the transpose operator and $i \in \{1, 2, ..., M\}$.

In addition to these dynamic quantities, we postulate also two memory locations encoded in the variables $p_n^{i,L}$, the *local* best of the ith particle, and p_n^g, the *global* best, both for the nth dimension.

The basic idea of the classical PSO algorithm is the clever exchange of information about the global and local best values mentioned above. Let us assume that the optimization goal is to maximize an objective function $f(\mathbf{r})$ in the manner described in Sec. 1.1. Each particle will examine its performance through the following two views.

1. **The Individual Perspective**: Here each particle will evaluate its performance (by evaluating the fitness function). At the lth iteration, the ith particle compares its present fitness $f(\mathbf{r}^i)$ value with the ones stored in $\mathbf{p}^{i,L}(l)$. If $f(\mathbf{r}^i(l+1)) > f(\mathbf{p}^{i,L}(l))$, then the algorithm sets $\mathbf{p}^{i,L}(l+1) = \mathbf{r}^i$. By this method, a record of the best achieved by an individual particle is kept for the velocity update (to be described subsequently).

2. **The Social Perspective**: Here the particle looks to the performance of the *entire* swarm instead of focusing on just the individual performance of a particle. The best performance of *all* the particles is stored in the global best $\mathbf{p}^g$. The fitness value at the current iteration is then compared to the one calculated at $\mathbf{p}^g$ and the algorithm sets $\mathbf{p}^g(l+1) = \mathbf{r}^i$ if $f(\mathbf{r}^i) > f(\mathbf{p}^g(l))$.

The key idea in the PSO method is how to combine the two different types of perspectives described above in order to predict the best update for the position in the next iteration. The idea, originally proposed in [2], was to change the velocity component in a manner such that the increments contributed by the social/individual perspective are directly proportional to the difference between

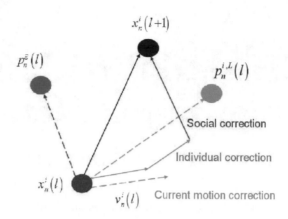

Figure 2.1: Graphical illustration of the mechanism of velocity update. At each iteration, the particle will combine information from the current motion (vector $\mathbf{v}^i(l)$, the social correction (vector $\mathbf{p}^{i,L}(l) - \mathbf{x}^i(l)$), and the individual correction (vector $\mathbf{p}^g(l) - \mathbf{x}^i(l)$). The contribution of the all these components will result in the prediction of the position vector in the next iteration as given by $\mathbf{x}^i(l+1)$.

the current position of the particle and the global/local best, respectively, which were recorded previously. This exchange is accomplished by the following two equations:

$$v_n^i(t + \Delta t) \;=\; w\,v_n^i(t) + c_1\varphi_1\left[p_n^{i,L} - x_n^i(t)\right]\Delta t + c_2\varphi_2\left[p_n^g - x_n^i(t)\right]\Delta t\;, \tag{2.3}$$

$$r_n^i(t + \Delta t) = r_n^i(t) + \Delta t\; v_n^i(t)\;, \tag{2.4}$$

where $n \in \{1, 2, ..., N\}$; Δt is the time step; c_1 and c_2 are the cognitive and social factors, respectively; φ_1 and φ_2 are two statistically independent random variables uniformly distributed between 0 and 1; w is the inertia factor.

Figure 2.1 demonstrates graphically the mechanism of position update. At each iteration l, the PSO algorithm combines three types of information in order to predict the best next position for the motion of the particles. These three information corresponds to the three terms in (2.3). The first term represents the contribution of the present motion in the overall decision about the next step. For the second information, the algorithm adds a vector correction [the second term in (2.3)] that accounts for the contribution of the individual knowledge accumulated during the evolution of the swarm. Finally, another vector correction [the third term in (2.3)] is added to include the contribution of the social knowledge.

Based on this procedure, the PSO algorithm will compare the objective function evaluated at the new positions with the error criterion set by the user as illustrated in Fig. 2.2. If the criterion is not satisfied, the random number generators in (2.3) will insure that different numerical values will

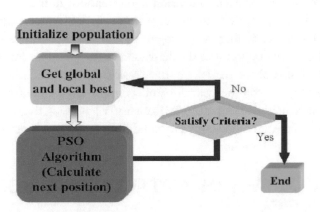

Figure 2.2: Flow chart for the PSO algorithm.

be tried in the next update and the process can go on until the termination of the evolution of the algorithm.

In the original algorithm [2], w was set to unity. The two random variables appearing in (2.3) are essential for the stochastic exploration of the search space and provide the source of the "randomness" in the PSO algorithm. It is noticed here that the two constants, c_1 and c_2, determine the relative weights of the individual and social perspectives, respectively, in the final decision made by the algorithm for updating the positions. There are different opinions in literature about how to choose the values of these constants [3]. However, in this book we will set all of the examples considered to the choice $c_1 = c_2 = 2.0$. It will be shown in Ch. 5 that one of the advantages of the quantum version of the PSO method is the elimination of these tuning parameters in a natural way.

Although the original version of the PSO started with the inertia weight $w = 1$, it turned out later that the performance of the algorithm can be dramatically improved if a variation in this tuning factor is introduced slowly during the evolution of the swarm. In most of the examples applied to benchmark functions and electromagnetic problems considered in this book, we choose to vary w linearly from 0.9 to a smaller value $w_{\min}$. Typical choices are $w_{\min} \in [0.2, 0.4]$. More details on other choices for the tuning of the PSO control parameters can be found in [3], [4], [5], [7], [9]. In Ch. 3, we provide a new insight on the physical meaning of the factor w and explain the difference in physics between the case of $w = 1$ and $w < 1$.

It has been shown in [4] that in order for the basic PSO algorithm ($w = 1$) to converge, all particles must approach the location $\mathbf{P}^i$ given by:

$$P_n^i = \frac{1}{\varphi_1 + \varphi_2} \left(\varphi_1 p_n^{i,L} + \varphi_2 p_n^{g} \right) , \tag{2.5}$$

where it is understood that for each dimension n the random number generators of φ_1 and φ_2 are initialized with different seeds. The convergence of the PSO algorithm to the location (2.5) can be guaranteed by proper tuning of the cognitive and social parameters of the algorithm [4]. Equation (2.5) will be very important in the derivation of the quantum PSO algorithm in Ch. 5.

To prevent explosion of the particles in the classical PSO, a maximum velocity V_{max} is introduced in each dimension to confine swarm members inside the boundary walls of the domain of interest. That is, if the velocity calculated by (2.3) satisfies $|V| > V_{max}$, then we simply set $|V| = V_{max}$ (see Fig. 4.2). Such choices are called boundary conditions and Ch. 4 is devoted to the study of this important aspect of the PSO method.

2.2   PARTICLE SWARM OPTIMIZATION AND ELECTRO-MAGNETICS

The literature on the theory and applications of the PSO algorithm is vast and no comprehensive coverage can be attempted in this book. Aside from developments in the fundamentals of the algorithm itself (e.g., see [2], [3], [4], [5]), numerous applications have been integrated successfully with the new algorithm in areas ranging from power systems, electronics, electromagnetics, to mechanical design and imaging in biological systems. However, it is conspicuous that applications to electromagnetics, and particularly antenna designs, have accounted for the lion share of the overall applications of the PSO method [36]. We provide below a very brief selective view of some of the important developments.

A general introduction to the PSO algorithm for electromagnetic applications can be found in [6], where the design of a magnetostatic system was considered. A more comprehensive review was given in [7], where the relative merits of the new optimization method was clearly highlighted in conjunction with the design of corrugated horn antenna problem. Also, in the same work, the concept of boundary conditions was investigated.

Applications of optimization methods to array antennas are popular because of the simplicity of the implementation of the objective function. Some work in this area employing the PSO method can be found in [8], [9], [20], [33], [35]. Extension of the PSO method to deal with multi-objective optimization problems was achieved in [18], [22], [23].

Comprehensive comparison with the genetic algorithm [37], [38], [39] was preformed in [9], [11]. For example, Fig. 2.3 illustrates a comparative study presented in [9] in which both of the PSO and the GA were utilized to perform the same design of linear phased-array antenna using amplitude-only, phase-only, and complex tapering. It is inferred from these figures that the performance of the PSO algorithm is competitive with the GA while the latter requires a much higher computational demand. Moreover, as can be seen from the general introduction in Sec. 2.1, the PSO algorithm requires much less number of tuning parameters compared with the GA method. This feature, together with the lower computational demand and implementation simplicity, are among the chief reasons of why the PSO is so popular in engineering electromagnetics.

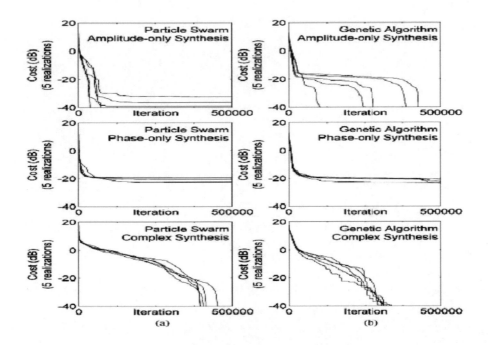

Figure 2.3: The results in [9] performance comparison between (a) particle swarm optimizer and (b) genetic algorithm; five trials each. (Reprinted here with permission from *IEEE Trans. Antennas Propagat.*, which is in [9].)

CHAPTER 3

Physical Formalism for Particle Swarm Optimization

3.1 INTRODUCTION

In this chapter, we propose an inter-disciplinary approach to particle swarm optimization (PSO) by establishing a molecular dynamics (MD) formulation of the algorithm, leading to a physical theory for the swarm environment. The physical theory provides new insights on the operational mechanism of the PSO method. In particular, a thermodynamic analysis, which is based on the MD formulation, can be introduced to provide deeper understanding of the convergence behavior of the basic classical PSO algorithm. The thermodynamic theory is used to propose a new acceleration technique for the PSO, which we apply to the problem of synthesis of linear array antennas.

Moreover, we will conduct a macroscopic study of the PSO to formulate a diffusion model for the swarm environment. Einstein's diffusion equation is solved for the corresponding probability density function (pdf) of the particles trajectory. The diffusion model for the classical PSO is used, in conjunction with Schrödinger's equation for the quantum PSO, to propose a generalized version of the PSO algorithm based on the theory of Markov chains. This unifies the two versions of the PSO, classical and quantum, by eliminating the velocity and introducing position-only update equations based on the probability law of the method.

3.2 MOLECULAR DYNAMICS FORMULATION

The main goal of molecular dynamics (MD) is to simulate the state of a system consisting of a very large number of molecules. A true consideration of such systems requires applying quantum-theoretic approach. However, the exact description of the quantum states requires solving the governing Schrödinger's equations, a task that is practically impossible for large many-particle systems [43]. Alternatively, MD provides a shortcut that can lead to accurate results in spite of the many approximations and simplifications implied in its procedure [45].

MD is based on calculating, at each time step, the positions and velocities, for all particles, by direct integration of the equations of motions. These equations are constructed basically from a classical Lagrangian formalism in which interactions between the particles are assumed to follow certain potential functions. The determination of the specific form of these potentials depends largely on phenomenological models and/or quantum-theoretic considerations.

Besides MD, there are in general three other methods to study the evolution of large number of particles: quantum mechanics, statistical mechanics, and Monte Carlo [45]. However, MD is preferred in our context over the other methods because it permits a direct exploitation of the structural similarities between the discrete form of the update equations of the PSO algorithm and the Lagrangian formalism.

3.2.1   Conservative PSO Environments

In order to formally construct the analogy between PSO and Newtonian mechanics, we consider a set of M identical particles, all with mass m, interacting with each other. We start first by a *conservative* system described by a Lagrangian function given by

$$L\left(r^i, \dot{r}^i\right) = \sum_{i=1}^{M} \frac{1}{2} m \, \dot{r}^i \cdot \dot{r}^i \; - \; U\left(r^1, r^2, \cdots, r^M\right) , \tag{3.1}$$

where r^i and $\dot{r}^i = v^i$ are the position and velocity of the ith particle, respectively. U is the potential function, which describes the intrinsic strength (energy) of the spatial locations of the particle in the space. The equations of motion can be found by searching for the critical value of the action integral

$$S = \int_{t_1}^{t_2} L\left(r^1, r^2, ..., r^M; \dot{r}^1, \dot{r}^2, ..., \dot{r}^M\right) dt , \tag{3.2}$$

where t_1 and t_2 are the initial and final times upon which the boundary of the trajectory is specified. The solution for this "optimization" problem is the Euler-Lagrange equation:

$$\frac{\partial L}{\partial x^i} - \frac{d}{dt} \frac{\partial L}{\partial \dot{x}^i} = 0 . \tag{3.3}$$

Equations (3.1) and (3.3) lead to

$$a^i = \dot{v}^i = \ddot{r}^i = \frac{F^i}{m} , \tag{3.4}$$

where F^i is the mechanical force acting on the ith particle in the swarm and a^i is its resulted acceleration. The mechanical force can be expressed in terms of the potential function U as follows:

$$F^i = -\frac{\partial}{\partial r^i} U\left(r^1, r^2, ..., r^M\right) . \tag{3.5}$$

Equations (3.1)–(3.5) represent a complete mechanical description of the particle swarm evolution; it is basically a system of continuous ordinary differential equations in time. To map this continuous version to a discrete one, like the PSO in (2.3) and (2.4), we consider the Euler-Cauchy discretization scheme [46]. Hence, it is possible to write the equations of motion in discrete time as

$$v\left(k\Delta t\right) = v\left((k-1)\,\Delta t\right) + \Delta t \, a\left(k\Delta t\right) \tag{3.6}$$

and

$$r\left(k\Delta t\right) = r\left((k-1)\,\Delta t\right) + \Delta t \, v\left(k\Delta t\right) , \tag{3.7}$$

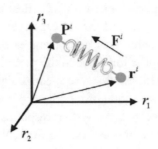

Figure 3.1: A 3-D illustration of the mechanical analogy to the PSO. The ith particle will experience a mechanical force identical to a spring with one end attached to the position and the other end attached to the particle's itself.

where Δt is the time step of integration.

By comparing Eqs. (2.3) and (2.4) to (3.6) and (3.7), it can be concluded that the PSO algorithm corresponds exactly to a swarm of classical particles interacting with a field of conservative force only if $w = 1$, which corresponds to the basic form of the PSO algorithm originally proposed in [2]. The acceleration is given by the following compact vector form:

$$a^i (k\Delta t) = \bar{\Phi} \left[P^i - r^i (k\Delta t) \right] , \tag{3.8}$$

where $\bar{\Phi}$ is a diagonal matrix with the nonzero elements drawn from a set of mutually exclusive random variables uniformly distributed from 0 to 1.

The integer $k \in \{1, 2, ..., N_{\text{itr}}\}$, where N_{itr} is the total number of iterations (generations), represents the time index of the current iteration. Thus, the force acting on the ith particle at the kth time step is given by

$$F^i (k\Delta t) = m\, \bar{\Phi} \left[P^i - r^i (k\Delta t) \right] . \tag{3.9}$$

Equation (3.9) can be interpreted as Hooke's law. In the PSO algorithm, particles appear to be driven by a force directly proportional to the displacement of their respective positions with respect to some center given by (2.5). Thus, for each particle there exists an equivalent mechanical spring with an anisotropic. Hooke's tensor equal to $m\bar{\Phi}$. Figure 3.1 illustrates this analogy.

The mass of the particle appearing in Eq. (3.9) can be considered as an extra parameter of the theory that can be chosen freely. This is because the basic scheme of the PSO method assumes point-like particles. Therefore, we will choose the mass m such that the simplest form of the quantities of interest can be obtained.

3.2.2 Philosophical Discussion

The main idea in the proposed connection between the PSO algorithm and the physical system descried by the Lagrangian (3.1) is to shift all of the 'randomness' and social intelligence (represented by p) to the law of force acting on particles in a way identical to Newton's law in its discrete-time form. This is as if there exists a hypothetical Grand Observer who is able to monitor the motion of all particles, calculates the global and local best, averages them in a way that appears random only to other observers, and then applies this force to the particles involved. Obviously, this violates relativity in which nothing can travel faster than the speed of light but Newtonian mechanics is not local in the relativistic sense. The assumption of intelligence is just the way the mechanical force is calculated at each time instant. After that, the system responds in its Newtonian way. There is no restriction in classical mechanics to be imposed on the force. What makes nature rich and diverse is the different ways in which the mechanical law of force manifests itself in Newton's formalism. In molecular dynamics (MD), it is the specific (phenomenological) law of force what makes the computation a good simulation of the quantum reality of atoms and molecules.

However, on a deeper level, even this Grand Observer who monitors the performance of the PSO algorithm can be described mathematically. That is, by combining Eqs. (3.5) and (3.9), we get the following differential equation for the potential U:

$$\frac{\partial U}{\partial \mathbf{r}^i} + m\, \bar{\bar{\mathbf{\Phi}}} \left(\mathbf{P}^i - \mathbf{r}^i \right) = 0 \,. \tag{3.10}$$

The reason why we did not attempt to solve this equation is the fact that P, as defined in (2.5), is a complicated function of the positions of all particles, which enforces on the problem the many-body interaction theme. Also, the equation cannot be solved uniquely since we know only the discrete values of P, while an interpolation/extrapolation to the continuous limit is implicit in the connection we are drawing between the PSO algorithm and the physical system. Moreover, since P is not a linear function of the interaction between particles, then Eq. (3.10) is a complex many-body nonlinear equation. While it might be solvable in principle, the nonlinearity may produce very complicated patterns that appears to be random (chaos) because of the sensitivity to the precision in the initial conditions. However, all of this does not rule out the formal analogy between the PSO and the Lagrangian physical system, at least on the qualitative level. In the remaining parts of this chapter, we will take U to represent some potential function with an explicit form that is not known to us, an assumption that does lead to serious modifications in the results presented here.

Regarding to the appearance of random number generators in the law of force (3.8), few remarks are in order. It should be clear that a Newtonian particle moves in a deterministic motion because of our *infinite*-precision knowledge of the external force acting on it, its mass, and the initial conditions. According to $F = ma$, if m is known precisely, but F is random (let us say because of some ignorance in the observer state), then the resulting trajectory will look random. However, and this is the key point, the particle does not cease to be Newtonian at all! What is Newtonian, anyhow, is the dynamical law of motion, not the superficial, observer-dependent, judgment that the motion looks random or deterministic with respect to *his* knowledge.

This can be employed to reflect philosophically on the nature of the term 'intelligence.' We think that any decision-making must involve some sort of ignorance or state of lack of knowledge. Otherwise, AI can be formulated as a purely computational system (eventually by a universal Turing machine). However, the controversial view that such automata can ultimately describe the human mind was evacuated from our discussion at the beginning by pushing the social intelligence P to the details of the external mechanical force law, and then following the consequences that can be derived by considering that the particles respond in a Newtonian fashion. We are fully aware that no proof about the nature of intelligence was given here in the rigorous mathematical sense, although something similar has been attempted in literature [48].

To summarize, the social and cognitive knowledge are buried in the displacement origin $\mathbf{P}^i$, from which the entire swarm will develop its intelligent behavior and search for the global optimum within the fitness landscape. The difficulty in providing full analysis of the PSO stems from the fact that the displacement center $\mathbf{P}^i$ is varying with each time step, and for every particle, according to the information on the entire swarm, rendering the PSO inherently a many-body problem in which each particle interacts with all the others.

3.2.3   Nonconservative (Dissipative) PSO Environments

It is important to state that for the transformation between the PSO and MD, as presented by Eq. (3.8), to be exact, the inertia factor w was assumed to be unity. However, as we pointed out in Sec. 3.2, for satisfactory performance of the PSO algorithm in real problems one usually reverts to the strategy of linearly decreasing w to a certain lower value. In this section, we will study in details the physical meaning of this linear variation.

The velocity equation for the PSO is given by:

$$\boldsymbol{v}^i\left(k\Delta t\right) = w\boldsymbol{v}^i\left((k-1)\,\Delta t\right) + \bar{\boldsymbol{\Phi}}\left[\boldsymbol{P}^i\left((k-1)\,\Delta t\right) - \boldsymbol{x}^i\left((k-1)\,\Delta t\right)\right]\Delta t\ . \tag{3.11}$$

Thus, by rearranging terms we can write:

$$\frac{\boldsymbol{v}^i(k\Delta t) - \boldsymbol{v}^i((k-1)\Delta t)}{\Delta t} = -\frac{1-w(t)}{\Delta t}\boldsymbol{v}^i\left((k-1)\,\Delta t\right) \\ + \bar{\boldsymbol{\Phi}}\left[\boldsymbol{P}^i\left((k-1)\,\Delta t\right) - \boldsymbol{x}^i\left((k-1)\,\Delta t\right)\right]\ . \tag{3.12}$$

Here $w = w(t)$ refers to a general function of time (linear variation is one common example in the PSO community). By taking the limit when $\Delta t \longrightarrow 0$, we find:

$$\boldsymbol{a}^i\left(t\right) = \beta\boldsymbol{v}^i\left(t\right) + \bar{\boldsymbol{\Phi}}\left[\boldsymbol{P}^i\left(t\right) - \boldsymbol{x}^i\left(t\right)\right]\ , \tag{3.13}$$

where

$$\beta = \lim_{\Delta t \to 0}\frac{w\left(t\right)-1}{\Delta t}\ . \tag{3.14}$$

Comparing Eq. (3.8) with (3.13), it is clear that the conservative Lagrangian in (3.1) cannot admit a derivation of the PSO equation of motion when w is different from unity; there exists a term proportional to velocity that does not fit with Newton's second law as stated in (3.4).

This problem can be solved by considering the physical meaning of the extra terms. For a typical variation of w starting at unity and ending at some smaller value (usually in the range 0.2–0.4, depending on the objective function), then we find from (3.14) that β is negative. That is, since the *total* force is given by the product of the acceleration $\mathbf{a}^i$ in (3.13) and the mass m, then it seems that the term $\beta \mathbf{v}^i$ counts for "friction" in the system that tends to lower the absolute value of the velocity as the particles evolve in time. In other words, w amounts to a *dissipation* in the system with strength given by the factor β. This explains why the conservative Lagrangian failed to produce the equations of motion in this particular case since the system is actually nonconservative.

Fortunately, it is still possible to employ a modified version of the Lagrangian formalism to include the dissipative case in our general analogy to physical systems. To accomplish this, we split the Lagrangian into two parts.

The first is L', which represents the conservative part and consists of the difference between kinetic and potential energies, as in (3.1), and is repeated here for convenience:

$$L' = \frac{1}{2} \sum_i m \, \dot{\boldsymbol{x}}^i \cdot \dot{\boldsymbol{x}}^i - U \, , \tag{3.15}$$

where the potential energy U is a function of the positions only.

The second part, L'', accounts for the nonconservative or dissipative contribution to the system. Following [49], we write:

$$L'' = \frac{1}{2} \sum_i \beta_i \dot{\boldsymbol{x}}^i \cdot \dot{\boldsymbol{x}}^i + \frac{1}{2} \sum_i \gamma_i \ddot{\boldsymbol{x}}^i \cdot \ddot{\boldsymbol{x}}^i \, . \tag{3.16}$$

Here, β_i represents the Rayleigh losses in the system and is similar to friction or viscosity. The second term containing γ_i accounts for radiation losses in the system. For example, if the particles are electrically charged, then any nonzero acceleration will force the particle to radiate an electromagnetic wave that carries away part of the total mechanical energy of the system, producing irreversible losses and therefore dissipation.

Both m and γ are assumed to be constants independent of time but this is not necessary for β_i. Moreover, the conservative Lagrangian L' has the dimensions of energy while the nonconservative part L'' has the units of power.

The equation of motion for the modified Lagrangian is given by [49]:

$$\left\{ \frac{\partial L''}{\partial \boldsymbol{x}^i} - \frac{d}{dt} \frac{\partial L''}{\partial \ddot{\boldsymbol{x}}^i} \right\} - \left\{ \frac{\partial L'}{\partial \boldsymbol{x}^i} - \frac{d}{dt} \frac{\partial L'}{\partial \dot{\boldsymbol{x}}^i} \right\} = 0 \, . \tag{3.17}$$

By substituting (3.15) and (3.16) to (3.17), we get:

$$-\gamma_i \frac{d}{dt} \ddot{\boldsymbol{x}}^i + m \ddot{\boldsymbol{x}}^i + \beta_i \dot{\boldsymbol{x}}^i = -\frac{\partial U}{\partial \boldsymbol{x}^i} \, . \tag{3.18}$$

Comparing (3.18) with (3.13), we immediately find:

$$\gamma_i = 0 \tag{3.19}$$

and

$$\beta_i = -m\beta = m \lim_{\Delta t \to 0} \frac{1 - w(t)}{\Delta t}, \qquad (3.20)$$

where Eq. (3.10) has been used.

The result (3.19) implies that there is no electromagnetic radiation in the PSO environment. One can say that the idealized particles in the PSO algorithm do not carry electric charge. However, it is always possible to modify the basic PSO to introduce new parameters. One of the main advantages of constructing a physical theory is to achieve an intuitive understanding of the meaning of tuning parameters in the algorithm. We suggest, for future work, considering the idea of adding an electric charge to the particles in the algorithm and investigate whether this may lead to an improvement in the performance or not.

Equation (3.20) amounts to a connection between the physical parameter β, the Rayleigh constant of friction, and a tuning parameter in the PSO algorithm, namely w. This motivates the idea behind slowly decreasing the inertia w during convergence. As $w(t)$ decreases, $\beta(t)$ increases, which means that the dissipation of the environment will increase leading to faster convergence. From this discussion we see that the common name attributed to w is slightly misleading. From the mechanical point of view, inertia is related to the mass m while w controls the dissipation or the friction of the system. A more convenient name for w would be something like the 'friction constant' or the 'dissipation constant' of the PSO algorithm.

The fact that the system is dissipative when w is less than unity may lead to some theoretical problems in the next sections, especially those related to thermodynamic equilibrium. However, a careful analysis of the relative magnitudes of the 'friction force' and the 'PSO force' in (3.18) shows that for a typical linear variation of w from 0.9 to 0.2, the friction becomes maximum at the final iteration, most probably when the algorithm has already converged to some result. At the beginning of the run, the friction is minimum while particles are expected to be 'reasonably' away from the local and global bests; that is, the difference $\varphi(x - p)$ is large compared to βv. This means that as the algorithm evolves to search for the nontrivial optima, the friction is not very significant. It becomes so only at the final iterations when β in (3.14) is already large compared to the PSO force (3.9). This explains why we have to be careful in choosing w; specifically, if the dissipation is increasing faster than the "intelligent" PSO force, immature convergence may occur and the algorithm will not be able to catch the global optimum. Based on this discussion, we will always assume that $w(t)$ is chosen to vary with time "wisely" (i.e., by ensuring that immature convergence is avoided). Therefore, we will ignore the effect of dissipation and treat the PSO as a collection of perfectly Newtonian particles interacting in conservative environment. Further results and studies in the remaining sections confirm this assumption.

3.3 EXTRACTION OF INFORMATION FROM SWARM DYNAMICS

The dynamic history of a swarm of particles is completely defined either by the system of Eqs. (2.3) and (2.4) for the PSO, or (3.1)–(3.8) for MD. The two representations are equivalent, where the transformation between one to the other is obtained by the mapping in (3.8). After finishing the simulation, we will end up with a set of M trajectories, each describing the successive velocities (momentums) and positions of one particle during the course of time. Any trajectory is assumed to be 1-D surface (curve) in an abstract $2MN$-dimensional vector space. We assume that this space forms a manifold through the adopted coordinate system and call it the phase space of the swarm system.

Let us write the trajectory of the ith particle as

$$\boldsymbol{\Gamma}_i(t) = \left\{ \left(\boldsymbol{r}^i(t), \boldsymbol{p}_m^i(t) \right), \forall t \in I_t \right\} \tag{3.21}$$

in the continuous form and

$$\boldsymbol{\Gamma}_i(k) = \left\{ \left(\boldsymbol{r}^i(k\Delta t), \boldsymbol{p}_m^i(k\Delta t) \right) \right\}_{k=1}^{N_{\text{itr}}} \tag{3.22}$$

for the discrete case. Here, I_t is the continuous time interval of the simulation and N_{itr} is the number of iterations. The *swarm dynamic history* of either the MD or the PSO can be defined as the set of all particles trajectories obtained after finishing the simulation

$$\bar{\boldsymbol{\Gamma}}(t) = \left\{ \boldsymbol{\Gamma}^i(t) \right\}_{i=1}^{M}. \tag{3.23}$$

One of the main objectives of this book is to provide new insights on how the classical PSO algorithm works by studying different quantities of interest. We define any dynamic observable of the swarm dynamics to be a smooth function of the swarm dynamic history $\bar{\boldsymbol{\Gamma}}(t)$. That is, the general form of a dynamic property (observable) will be written as

$$A_\Phi(t) = \Phi \bar{\boldsymbol{\Gamma}}(t), \tag{3.24}$$

where Φ is a sufficiently smooth operator.

The PSO is inherently a stochastic process as can be inferred from the basic Eqs. (2.3) and (2.4). Moreover, we will show later that a true description of MD must rely on statistical mechanical considerations where the particle's trajectory is postulated as a stochastic process. Therefore, single evaluation of a dynamic property, as defined in (3.24), cannot reflect the qualitative performance of interest. Instead, some sort of averaging is required in order to get reasonable results. We introduce here two types of averages. The first one is the *time* average defined as:

$$\langle A_\Phi \rangle = \lim_{T \to \infty} \frac{1}{T} \int_{t_0}^{t_0+T} A_\Phi(t)\, dt \tag{3.25}$$

and

$$\langle A_\Phi \rangle = \frac{1}{N_{\text{itr}}} \sum_{k=1}^{N_{\text{itr}}} A_\Phi(k) \qquad (3.26)$$

for continues and discrete systems, respectively. An important assumption usually invoked in MD simulations is that an average over finite interval should estimate the infinite integration/summation of (3.26) [43]. Obviously, this is because any actual simulation interval must be finite. The validity of this assumption can be justified on the basis that the number of time steps in MD simulations is large. However, no such restriction can be imposed in the PSO solutions because in engineering applications, especially real-time systems, the number of iterations must be kept as small as possible to facilitate good performance in short time intervals. Therefore, instead of time averages, we will invoke the *ensemble* average, originally introduced in statistical mechanics.

We denote the ensemble average, or the expected value, of a dynamic observable A_Φ as $E[A_\Phi]$. It is based on the fact that our phase space is endowed with a probability measure so meaningful probabilities can be assigned to outcomes of experiments performed in this space. Moreover, if the system that produced the swarm dynamics in (3.23) is ergodic, then we can equate the time average with ensemble average. In other words, under the ergodic hypothesis we can write [43], [47]:

$$\langle A_\Phi \rangle = E[A_\Phi] . \qquad (3.27)$$

Therefore, by assuming that the ergodicity hypothesis is satisfied, one can always perform ensemble average whenever a time average is required. Such an assumption is widely used in MD and statistical physics [43], [45]. However, in the remaining parts of this book we employ ensemble average, avoiding therefore the controversial opinion of whether all versions of the PSO algorithm are strictly ergodic or not. Moreover, the use of ensemble average is more convenient when the system under consideration is dissipative.

3.4 THERMODYNAMIC ANALYSIS OF THE PSO ENVIRONMENT

3.4.1 Thermal Equilibrium

Thermal equilibrium can be defined as the state of a system of molecules in which no change of the macroscopic quantities occurs with time. This means that either all molecules have reached a complete halt status (absolute zero temperature), or that averages of the macroscopic variables of interest do not change with time. To start studying the thermal equilibrium of a swarm of particles, we will employ the previous formal analogy between the PSO and classical Newtonian particle environments to define some quantities of interest, which will characterize and enhance our understanding of the basic mechanism behind the PSO.

In physics, the temperature of a group of interacting particles is defined as the average of the kinetic energies of all particles. Based on kinetic theory, the following expression for the temperature can be used [43]:

$$T\left(t\right) = \frac{1}{M} \frac{m}{N k_B} \sum_{i=1}^{M} |\boldsymbol{v}_i\left(t\right)|^2 \, , \tag{3.28}$$

where k_B is Boltzmann's constant and for the conventional 3-D space we have $N = 3$. Analogously, we define the particle swarm temperature as

$$T\left(t\right) = E\left[\frac{1}{M} \sum_{i=1}^{M} |\boldsymbol{v}_i\left(t\right)|^2\right] \, , \tag{3.29}$$

where without any loss of generality the particle's mass is assumed to be

$$m = N k_B \, . \tag{3.30}$$

Since there is no exact *a priori* knowledge of the statistical distribution of particles at the off-equilibrium state, the expected value in (3.29) must be estimated numerically. We employ the following estimation here

$$T\left(t\right) = \frac{1}{M B} \sum_{j=1}^{B} \sum_{i=1}^{M} \left|\boldsymbol{v}_i^j\left(t\right)\right|^2 \, , \tag{3.31}$$

where B is the number of repeated experiment. In the jth run, the velocities $\boldsymbol{v}_i^j\left(t\right), i = 1, 2, .., M,$ are recorded and added to the sum. The total result is then averaged over the number of runs. The initial positions and velocities of the particles in the PSO usually follow a uniformly random distribution. At the beginning of the PSO simulation, the particles positions and velocities are assigned random values. From the kinetic theory of gases point of view, this means that the initial swarm is not at the thermal equilibrium. The reason is that for thermal equilibrium to occur, particles velocities should be distributed according to Maxwell's distribution [43], which is, strictly speaking, a Gaussian probability distribution. However, since there is in general no prior knowledge about the location of the optimum solution, it is customary to employ the uniformly random initialization.

In general, monitoring how the temperature evolves is not enough to determine if the system had reached an equilibrium state. For isolated systems, a sufficient condition to decide whether the system has reached macroscopically the thermal equilibrium state is to achieve maximum entropy. However, the calculation of entropy is difficult in MD [43]. Usually, MD simulations produce automatically many macroscopic quantities of interest; by monitoring all of them, it is possible to decide whether the system had reached equilibrium or not. However, in the basic PSO algorithm there are no corresponding quantities of interest. What we need is really one or two auxiliary quantities that can decide, with high reliability, whether the system has converged to the steady thermal state or not. Fortunately, such a measure is available in literature in a form called the α-

factor. We employ the following thermal index [44]:

$$\alpha(t) = \frac{1}{B} \sum_{j=1}^{B} \frac{\frac{1}{M} \sum_{i=1}^{M} \left\| v_i^j \right\|^4}{\left[\frac{1}{M} \sum_{i=1}^{M} \left\| v_i^j \right\|^2 \right]^2}, \tag{3.32}$$

where the usual definition of the Euclidean norm for a vector $\mathbf{v}$ with length N is given by

$$\| v \| = \sqrt{\sum_{n=1}^{N} v_n^2}. \tag{3.33}$$

Here, v_i^j is the velocity of the ith particle at the jth experiment. At equilibrium, the index above should be around 5/3 at isothermal equilibrium [44].

3.4.2 Primary Study Using Benchmark Test Functions

In order to study the qualitative behavior of the swarm thermodynamics, we consider the problem of finding the global minimum of the following N-dimensional standard test functions:

$$f_1(x) = \sum_{n=1}^{N} x_n^2, \tag{3.34}$$

$$f_2(x) = \sum_{n=1}^{N} \left[x_n^2 - 10 \cos(2\pi x_n) + 10 \right], \tag{3.35}$$

$$f_3(x) = \sum_{n=1}^{N-1} \left[100 \left(x_{n+1} - x_n^2 \right)^2 + (x_n - 1)^2 \right]. \tag{3.36}$$

The sphere function in (3.34) has a single minimum located at the origin. The function defined in (3.35) is known as Rastigrin function with a global minimum located at the origin. This is a hard multimodal optimization problem because the global minimum is surrounded by a large number of local minima. Therefore, reaching the global peak without getting stuck at one of these local minima is extremely difficult. The problem in (3.36), known as Rosenbrock function, is characterized by a long narrow valley in its landscape with global minimum at the location $[1, 1, ..., 1]^T$. The value of the global minimum in all of the previous functions is zero.

Figure 3.2 shows the cost evolution together with the time history of the swarm temperature obtained from the PSO simulation of the three test functions (3.34)–(3.36). A swarm of 20 particles searching in a 10-dimensional space is considered for each case. Figure 3.2(a) indicates that only the sphere function has successfully converged to the global optimum. On the other hand, by examining the thermal behavior of the system as indicated by Fig. 3.2(b), it is clear that the swarm temperature drops rapidly, suggesting that the swarm is "cooling" while evolving with time. Eventually, the temperature goes to zero, indicating that the swarm has reached the state of thermal equilibrium. Notice that in thermodynamics, zero temperature is only a special case. Convergence to a constant nonzero temperature can also be characterized as thermal equilibriums.

It is interesting to observe from Fig. 3.2 that the thermal behaviors of the three solutions look indistinguishable. This is the case although the convergence curves of Fig. 3.2(a) shows that only the sphere function has converged to its global optimum. This demonstrates that convergence to thermal equilibrium does not mean that convergence to the global optimum has been achieved. Rather, convergence to a local optimum will lead to thermal equilibrium. The swarm temperature cannot be used then to judge the success or failure of the PSO algorithm.

The conclusion above seems to be in direct contrast to the results presented in [50]. This work introduced a quantity called average velocity defined as the average of the *absolute* values of the velocities in the PSO. Although the authors in [50] did not present a physical analogy between MD and PSO in the way we provided in Sec. 3.2, the average velocity will qualitatively reflect the behavior of the swarm temperature as defined in (3.31). The main observation in [50] was that convergence of the 'average velocity' (temperature) to zero indicates global convergence. Based on this, they suggested an adaptive algorithm that forces the PSO to decrease the average velocity in order to guarantee successful convergence to global optimum. However, it is clear from the results of Fig. 3.2 that this strategy is incorrect. Actually, it may cause the algorithm to be trapped in a local optimum.

A better perspective on the thermal behavior of the system can be obtained by studying the evolution of the α-index. Figure 3.3 illustrates an average of 1,000 run of the experiments of Fig. 3.2. Careful study of the results reveals a different conclusion compared with that obtained from Fig. 3.2(b). It is clear from Fig. 3.2(a) that the PSO solutions applied to the three different functions converge to a steady-state value at different number of iterations. The sphere function, Rastigrin function, and Rosenbrock valley converge roughly at around 400, 300, and 350 iterations, respectively. By examining the time histories of the α-index in Fig. 3.3, it is seen that the sphere function's index is the fastest in approaching its theoretical limit at the number of iterations where convergence of the cost function has been achieved. For example, the thermal index of the Rastigrin case continued to rise after the 300th iteration until the 350th iteration where it starts to fall. Based on numerous other experiments conducted by the authors, it seems more likely that the α-index can be considered an indicator to decide if the PSO algorithm has converged to the global optimum or not. It seems that convergence to global optimum in general corresponds to the fastest convergence to thermal equilibrium as indicated by our studies of the associated α-index. Moreover, it is even possible to consider that the return of the α-index to some steady state value, i.e., forming a peak, as a signal that convergence has been achieved. However, we should warn the reader that more comprehensive work with many other objective functions is required to study the role of the α-index as a candidate for global optimum convergence criterion.

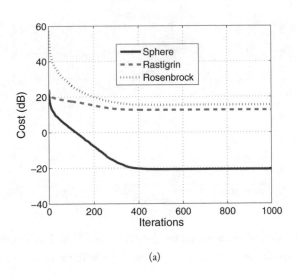

(a)

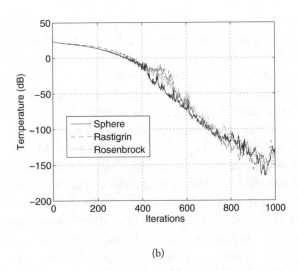

(b)

Figure 3.2: Cost function and temperature evolution for a swarm of 20 particles searching for the minimum in a 10-dimensional space. The parameters of the PSO algorithm are $c_1 = c_2 = 2.0$, and w is decreased linearly from 0.9 to 0.2. The search space for all the dimensions is the interval [-10,10]. A maximum velocity clipping criterion was used with $V_{max} = 10$. The algorithm was run 100 times and averaged results are reported as: (a) Convergence curves; (b) temperature evolution. (Reprinted here with permission from *Progr. Electromagn. Wave Res.*, which is in [68].)

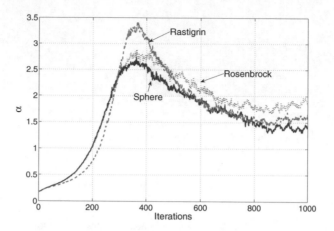

Figure 3.3: The time evolution of the α-index for the problem of Fig. 3.2. The PSO algorithm was run 1,000 times and averaged results are reported. (Reprinted here with permission from *Progr. Electromagn. Wave Res.*, which is in [68].)

The fact that the α-index is not converging exactly to 5/3 can be attributed to three factors. First, a very large number of particles is usually required in MD simulations to guarantee that averaged results will correspond to their theoretical values. Considering that to reduce the optimization cost we usually run the PSO algorithm with small number of particles, the obtained curves for the α-index will have noticeable fluctuations, causing deviations from the theoretical limit. Second, it was pointed out in Sec. 3.3 that because of the presence of an inertia factor w in the basic PSO equations with values different from unity, the analogy between the PSO and MD is not exact. These small deviations from the ideal Newtonian case will slightly change the theoretical limit of the α-index. Our experience indicates that the new value of α at thermal equilibrium is around 1.5. Third, the theoretical value of 5/3 for the α-index is based on the typical Maxwell's distribution at the final thermal equilibrium state. However, the derivation of this distribution assumes weakly interacting particles in the thermodynamic limit. It is evident, however, from the discussion of Sec. 3.2 that the PSO force is many-body global interaction that does not resemble the situation encountered with rarefied gases, in which the kinetic theory of Maxwell applies very well. This observation forces us to be careful in drawing quantitative conclusions based on the formal analogy between the PSO and the physical system of interacting particles.

3.4.3   Energy Consideration

Beside the basic update Eqs. (2.3) and (2.4), (3.6), and (3.7), it should be clear how to specify the particles behavior when they hit the boundaries of the swarm domain. Assume that the PSO system under consideration is conservative. If we start with a swarm total energy E, the law of conservation of energy states that E will stay constant as long as no energy exchange is allowed at the boundary. From the MD point of view, there are two possible types of boundary conditions (BC), dissipative and nondissipative. Dissipative boundary conditions refer to the situation when the special treatment of the particles hitting the wall leads to a drop in the total energy of the swarm. An example of this type is the absorbing boundary condition (ABC), in which the velocity of a particle hitting the wall is set to zero [43], [45]. This means that the kinetic energy of a particle touching one of the forbidden walls is instantaneously "killed." If a large number of particles hit the wall, which is the case in many practical problems, there might be a considerable loss of energy in the total swarm, which in turns limits the capabilities of the PSO to find the optimum solution for several problems. The reflection boundary condition (RBC) is an example of nondissipative BCs. However, for this to hold true reflections must be perfectly elastic; that is, when a particle hits the wall, only the sign of its vector velocity is reversed [43], [45]. This will keep the kinetic energy of the particle unchanged, leading to a conservation of the total energy of the swarm.

In the latter case, it is interesting to have a closer look to the energy balance of the PSO. From Eq. (3.1), we consider the total mechanical energy

$$E = K + U ,$$ (3.37)

where K and U are the total kinetic and potential energies, respectively. It is possible to show that this energy is always constant for the conservative Lagrangian defined in (3.1) [49].

According to the thermodynamic analysis of Sec. 3.4.1, when particles converge to the global optimum, the temperature drops rapidly. This means that while the swarm is evolving, its kinetic energy decreases. Since the total energy is conserved, this means that convergence in the PSO can be interpreted as a process in which the kinetic energy is continually converted to potential energy. The final thermodynamic state of the swarm at the global optimum represents, therefore, a unique spatial configuration in which the particles achieve the highest possible potential energy in the searched region in the configuration space. This conclusion applies only approximately to the case when the boundary condition is dissipative since the energy loss at the wall is small compared to the total energy of the swarm (provided that large number of particles is used).

3.4.4   Dynamic Properties

Autocorrelation functions measure the relations between the values of certain property at different times. We define here a collective velocity autocorrelation function in the following way:

$$\Psi\left(t\right) = \frac{\left\langle \frac{1}{M} \sum_{i=1}^{M} \boldsymbol{v}_i\left(t_0\right) \cdot \boldsymbol{v}_i\left(t_0 + t\right) \right\rangle}{\left\langle \frac{1}{M} \sum_{i=1}^{M} \boldsymbol{v}_i\left(t_0\right) \cdot \boldsymbol{v}_i\left(t_0\right) \right\rangle}$$ (3.38)

where t_0 is the time reference. The time average in (**??**) can be estimated by invoking the ergodicity hypothesis in the expected value operator in (3.24). The final expression will be:

$$\Psi(t) = \frac{1}{B} \sum_{j=1}^{B} \left\{ \frac{\sum_{i=1}^{M} \boldsymbol{v}_i^j(t_0) \cdot \boldsymbol{v}_i^j(t_0 + t)}{\sum_{i=1}^{M} \boldsymbol{v}_i^j(t_0) \cdot \boldsymbol{v}_i^j(t_0)} \right\}, \tag{3.39}$$

where B is the number of experiments.

Figure 3.4(a) illustrates the calculations of the autocorrelation function in (3.39) with the time reference $t_0 = 0$. The behavior of this curve has a very strong resemblance to curves usually obtained through MD simulations. Such curves may provide a lot of information about the underlying dynamics of the swarm environment. First, at the beginning of the simulation there are strong interactions between particles, indicating that the velocity of each particle will change as compared to the reference point. This is because, according to Newton's first law, if there are no forces acting on a particle the velocity will stay the same. However, in solids each molecule is attached to a certain spatial location. For this, it is hard to move a molecule far away from its fixed position. What happens then is that the particle will oscillate around its initial position. This explains the strong oscillations of $\Psi(t)$ in Fig. 3.4(a), which suggest that the PSO environment resembles a solid state. The oscillations will not be of equal magnitude, however, but decay in time, because there are still perturbative forces acting on the particles to disrupt the perfection of their oscillatory motion. So what we see is a function resembling a damped harmonic motion.

In liquids, because there are no fixed positions for the molecules, oscillations cannot be generated. Thus, what should be obtained is a dammed sinusoid with one minimum. For gases, all what can be observed is a strong decay following an exponential law.

The observation that for the three different cost functions we get very similar early time response reflects probably that the randomly uniform initialization of the PSO algorithm construct a similar crystal-like solid structures, making early interactions of the particles similar. However, after 20 iterations the responses become different, as we expect, since particles start to develop different inter-molecular forces according to the fitness landscape encountered.

Another way to look into the dynamics of the PSO is to calculate the Fourier spectrum of the time autocorrelation function as defined in (**??**). This gives some information about the underlying resonant frequencies of the structure. Figure 3.4(b) illustrates the calculated spectrum. The shape looks very similar to results obtained for solid materials in MD [45] where a profound single peak is apparent. The Fourier transform of (**??**) is also called the generalized phonon density of state [45].

We conducted numerous experiments with several test functions where it seems that the classification of the PSO environment into gas, liquid, solid is strongly dependent on the boundary conditions, population size, and number of dimensions. Table 3.1 summarizes the shape of the velocity autocorrelation function corresponding to the three cases.

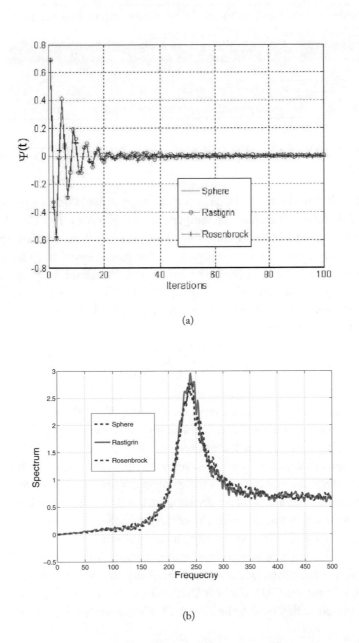

(a)

(b)

Figure 3.4: The velocity autocorrelation function for the problem of Fig. 3.2 with $B = 100$. (a) Time domain results; (b) frequency domain results with frequency given in terms of $1/\Delta t$. (Reprinted here with permission from *Progr. Electromagn. Wave Res.*, which is in [68].)

Table 3.1: Velocity autocorrelation function corresponding to different states for the PSO environment.

State	Velocity Autocorrelation
Gas	Damped exponential
Liquid	One period damped sinusoid with single minimum
Solid	Damped sinusoid with possibly multiple-oscillations

3.5   ACCELERATION TECHNIQUE FOR THE PSO ALGORITHM

In this section, we will provide an application for the thermodynamic analysis of Sec. 3.4. The idea is to employ the natural definition of temperature, as given by Eq. (3.31), in a technique similar to optimization using simulated annealing. The idea is the observation that the decay of temperature is a necessary condition for convergence to the global optimum. This means that if the PSO has enough recourse to search for the best solution, then temperature decay may be enhanced by artificially controlling the thermal behavior of the system. We employ a technique used in some MD simulations to push the swarm to equilibrium. Mainly, the velocity of each particle is modified according to the following formula:

$$\boldsymbol{v}_i(k+1) = \sqrt{\frac{T_0}{T(k)}} \boldsymbol{v}_i(k) \ , \tag{3.40}$$

where T_0 is a user-defined target temperature. The meaning of (3.40) is that while the swarm is moving toward the thermal equilibrium of the global optimum, the velocities are pushed artificially to produce a swarm with a pre-specified target temperature. In simulated annealing, the artificial cooling of the algorithm is obtained by a rather more complicated method where the probability is calculated for each iteration to decide how to update the particles. Moreover, there the biggest obstacle is the appearance of new control parameters, the temperature, which has no clear meaning. This exerts burden on the user while choosing the suitable value for the tuning parameters. However, it seems that the technique in (3.40) may provide an easier alternative.

The proposed acceleration technique will be demonstrated with a practical application from electromagnetics. We consider the design of linear array antenna to meet certain pre-defined goal given by a user-defined function $T(t)$. Suppose that we have N-element linear array separated by a uniform distance d (taken in this book to be $d = \lambda/2$). The normalized array factor is given by:

$$AF(u) = \frac{1}{AF_{\max}} \sum_{n=1}^{N} I_n e^{j2\pi n \, du/\lambda} \ , \tag{3.41}$$

where I_n are the amplitude coefficients (generally complex), and $u = \sin\theta$, where θ is the observation angle measured with respect to the normal to the array axis. AF stands for the array factor and $AF_{\max}$

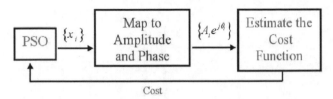

Figure 3.5: Block diagram representation of the linear array synthesis problem. Here the particles are $x_i \in \{0 \le x_i \le 1, i = 1, 2, ..., 2N\}$. Amplitudes and phases are ranged as $A_i \in \{A_{\min} \le A_i \le A_{\max}, i = 1, 2, ..., N\}$ and $\theta_i \in \{0 \le \theta_i \le 2\pi, i = 1, 2, ..., N\}$, respectively. (Reprinted here with permission from *Progr. Electromagn. Wave Res.*, which is in [68].)

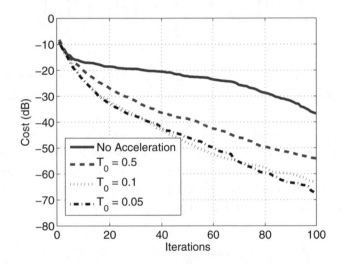

Figure 3.6: Convergence curves for the synthesis of symmetric pattern by using a linear array of 14 elements. The population size is 8 particles. The inertia factor w is decreased linearly from 0.9 to 0.2. $c_1 = c_2 = 2.0$. (Reprinted here with permission from *Progr. Electromagn. Wave Res.*, which is in [68].)

is the maximum value of the magnitude of the array factor defined as $AF_{\max} = \max\{|AF(u)|\}$, where the maximization operator is carried over the observational space u.

Figure 3.5 shows a block diagram representation of the linear array antenna synthesis problem. The algorithm, whether classical or quantum, works on particles moving in N-dimensional hyperspace. The information about the amplitudes and phases are encoded in the coordinates of the particle that are normalized between 0 and 1. The mapping block shown in Fig. 3.5 gives the required transformation between the abstract optimizer and the real physical problem.

The "Don't Exceed Criterion" is utilized in the formulation of the objective function. That is, an error will be reported only if the obtained array factor exceeds the desired sidelobe level. We start

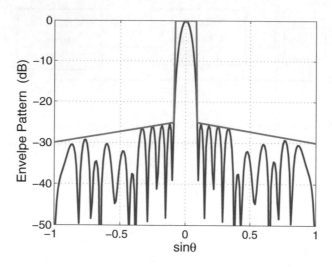

Figure 3.7: Radiation pattern of the linear array antenna corresponding to best case of Fig. 3.6 ($T_0 =$ 0.05). (Reprinted here with permission from *Progr. Electromagn. Wave Res.*, which is in [68].)

by generating M observation points in the sine space $u = \sin \theta$. Then, at each point we compare between the obtained array factor AF and T. If $|AF(u_k)|$ exceeds $T(u)$, then we store the error $e(u_k) = |AF(u_k)| - T(u_k)$. We end up with a vector of errors $\mathbf{e}$ that has nonzero positive values only at the locations where the obtained array factor exceeds the desired level. The cost function F is defined then as the averaged sum of the squares of the errors obtained and is expressed as:

$$F = \frac{1}{M} \sum_{k=1}^{M} [\max(|AF(u_k)| - T(u_k), 0)]^2 , \tag{3.42}$$

where M indicates the number of sampling points and $\Delta u = 2/(M-1)$.

The PSO algorithm is applied to synthesize an array antenna with a tapered side lobe mask that decreases linearly from -25 dB to -30 dB. The beamwidth of the array pattern is 9°. The total number of sampling points is 201. Figure 3.6 illustrates the convergence curves results where it is clear that the proposed artificial cooling can successfully accelerate the convergence of the method. The final obtained radiation pattern for $T_0 = 0.05$ is shown in Fig. 3.7.

3.6   DIFFUSION MODEL FOR THE PSO ALGORITHM

The development of the PSO algorithm presented so far was based on drawing a formal analogy between MD and the swarm environment. This means that a *microscopic* physical treatment is employed in modeling the PSO. In this section we develop an alternative view that is based on *macroscopic* approach. This new analogy can be constructed by postulating that particles in the

PSO environment follow a diffusion model. The use of diffusion models is now central in solid-state physics and molecular dynamics since the pioneering works of Boltzmann and Einstein more than one hundred year ago. Moreover, since Einstein's work on the origin of Brownian motion, the diffusion equation is used routinely as a theoretical basis for the description of the underlying stochastic dynamics of molecular processes.

The justification of the Diffusion model can be stated briefly as follows. From (3.13), we can write the equation of motion of a particle in the PSO algorithm as:

$$\ddot{\boldsymbol{x}}(t) = \beta \boldsymbol{v}(t) + \boldsymbol{\Gamma}(t), \tag{3.43}$$

where $\boldsymbol{\Gamma}(t) = \bar{\boldsymbol{\Phi}} \cdot [\boldsymbol{P}(t) - \boldsymbol{x}(t)]$. From the definition of $\boldsymbol{P}$ in (2.5), one can see that even in the continuous limit $\Delta t \to 0$ the term $\boldsymbol{\Gamma}(t)$ may be discontinuous because of the sudden update of the global and local positions. Thus, $\boldsymbol{\Gamma}(t)$ can be easily interpreted as a random fluctuations term that arises in the PSO environment in a way qualitatively similar to random collision. Notice that the particles in the PSO are originally thought of as soft spheres (i.e., there is no model for physical contacts between them leading to collisions). This interpretation brings the PSO immediately to the realm of the Brownian motion, where the dynamics is described by an equation similar to (3.43).

We start from the basic diffusion equation [43]:

$$\frac{\partial \tilde{N}(\boldsymbol{x}, t)}{\partial t} = D \nabla^2 \tilde{N}(\boldsymbol{x}, t), \tag{3.44}$$

where $\tilde{N}$ is the concentration function of the molecules, $\mathbf{x}$ is the position of the particle, and D is the diffusion constant.

Einstein obtained (3.44) in his 1905 paper on Brownian motion [52]. Direct utilization of this equation leads to solutions for the concentration of particles as functions of time and space. However, a revolutionary step taken by Einstein was to give the particle concentration different interpretations, deviating strongly from the classical treatment. To illustrate this, let us consider a 1-D diffusion equation as given below

$$\frac{\partial \tilde{N}}{\partial t} = D \frac{\partial^2 \tilde{N}}{\partial x^2}. \tag{3.45}$$

Einstein's interpretation of the physical meaning of the concentration function, $\tilde{N}$, started by considering a short time interval, Δt, in which the position changes from x to $x + dx$. A very important step was to refer each particle to its center of gravity [52]. Here, we propose to refer each particle to the location p_n^i, as defined in (2.5), where we consider this to be the *natural* center of gravity for particles in the PSO environment. Therefore, it is possible to define the new position variable

$$r_k^{l,n} = x_k^{l,n} - p_{k-1}^{l,n}, \tag{3.46}$$

where we have modified the notation to allow for the iteration index k to appear as the subscript in (3.46) while the indices of the particle number i and the dimension n are the superscripts.

Clearly, the diffusion Eq. (3.45) is invariant under the linear transformation (3.46). During the short time interval Δt, particle locations will change according to a specific probability low. This means that particles trajectories are not continuous deterministic motions, but a stochastic path reflecting the irregular motion of particles in noisy environments. Therefore, the re-interpretation of the diffusion equation in (3.44) is that it gives us the probability density function (pdf) of the motion, not the classical density. Assuming the following boundary conditions:

$$\tilde{N}(r, 0) = \delta(r), \quad \lim_{r \to \infty} \tilde{N}(r, t) = 0, \quad \int_{-\infty}^{\infty} \tilde{N}(r, t) \, dx = M \,, \tag{3.47}$$

the solution of (3.45) can be written as:

$$\tilde{N}(r, t) = \frac{M}{\sqrt{4\pi D}} \frac{e^{-\frac{r^2}{4Dt}}}{\sqrt{t}} \,, \tag{3.48}$$

where this can be easily verified by substituting (3.48) back into (3.45).

Equation (3.48), which represents a Gaussian distribution for the particle's positions, was observed empirically before [53]. The development of this section presents, therefore, a theoretical derivation for the pdf of particles moving in the PSO algorithms that matches the experimentally observed results. However, the standard interpretation of (3.48) in the theory of diffusion equations is that it represents the Green's function of the medium. That is, depending on what is the boundary-condition imposed on the initial swarm different distributions will result.

It is very important for the subsequent developments of this chapter to recognize that (3.48) should be interpreted as a *conditional* pdf. This can be inferred from the fact that the time variable appearing in (3.48) is referenced to the origin. As we mentioned above, about Einstein's analysis of the particles trajectory, the total motion can be expressed as a succession of short intervals of time. This means that if we start at a time instant t_{k-1}, then the conditional pdf can be expressed as:

$$f\left(x_k^{i,n}, \; t_k | x_{k-1}^{i,n}, t_{k-1}\right) = \frac{1}{\sqrt{4\pi D (t_k - t_{k-1})}} \exp\left\{-\frac{\left(x_k^{i,n} - p_{k-1}^{i,n}\right)^2}{4D (t_k - t_{k-1})}\right\} \,, \tag{3.49}$$

where f stands now for the pdf of the particles position. This final form is valid at arbitrary time instants.

It remains to see how the diffusion constant in (3.49) can be estimated. Here, we come to the main contribution of Einstein's paper in which he introduced the first connection between the macroscopic representation of the diffusion model and the microscopic representation of MD. Basically, the following relation was derived in [52]:

$$D = \lim_{t \to \infty} \frac{\langle [r(t) - r(0)]^2 \rangle}{6t} \,. \tag{3.50}$$

Clearly, this expression can be easily evaluated from the MD or PSO simulation. Therefore, it is possible to get the conditional pdf of the particles motion by calculating the diffusion constant as given in (3.50). It should be mentioned that the pdf is, in general, different for varying boundary conditions. The well-established theory of diffusion equations can be employed in solving for the exact pdf of PSO under these different boundary modes.

Few general remarks on the diffusion approach to the PSO are needed here. First, as we have just mentioned, the diffusion equation is subject to various boundary conditions that govern the constraints on the swarm evolution. The Gaussian pdf in (3.48) was stated only as a particular solution under some general conditions (3.47). oIf the PSO can be modeled as an evolution of particles under certain boundary modes, then one needs to construct the appropriate diffusion model and solve the boundary value problem. The formal implementation of this program requires further theoretical investigations and numerical studies, which are beyond the scope of the current work. The theory of diffusion is very rich and mature in theoretical physics; our sole purpose is to invite mathematically oriented researchers in the optimization community to consider learning from physicists in that regard.

3.7   MARKOV MODEL FOR SWARM OPTIMIZATION TECHNIQUES

3.7.1   Introduction

In this section, a unified approach to the PSO techniques presented in this book will be proposed based on the theory of Markov chains. One can visualize the Markov process as a hidden (deeper) layer in the description of dynamic problems. It does not present new behavior in the phenomenon under study, but may provide better insight on the way we write down the equations of motion. Since algorithms are sensitive to the various mathematical schemes used to implement them (e.g., some models are computationally more efficient), it is possible to gain new advantages, theoretical and conceptual, by investigating other deeper layers like Markov chains.

We start from the observation that the Gaussian pdf of particle positions lead to the possibility of eliminating the velocity from the algorithm and developing, instead, position-only update equations [53]. The recently introduced quantum version of the PSO has naturally no velocity terms [40], [41], [42]. This is because in quantum mechanics, according to Heisenberg's principle of uncertainty, it is not possible to determine both the position and the velocity with the same precision. It can be shown that all quantum mechanical calculations will produce a natural conditional pdf for the positions in the swarm environment as solutions of the governing Schrödinger's equation [42]. Therefore, one way to unify both the quantum and classical approaches is to consider their common theme: the existence of certain conditional probability distributions underlying the stochastic dynamics of the problem. Since this pdf can be obtained either by solving the diffusion equation for the classical PSO, or Schrödinger's equation for the quantum PSO, then the update equation can be directly derived from the pdf without using alluding to other physical arguments. An example will be given in the next section.

3.7.2 Derivation of the Update Equations Using Probability Theory

The diffusion equation and Schrödinger's equation can be considered as *sources* for various probability density functions that govern the stochastic behavior of moving particles. Now, since the pdf of the particle at the kth iteration is known, we will simulate the trajectory by transforming a uniformly distributed random variable into another random variable with the desired pdf. Let the required transformation that can accomplish this be denoted as $y = h(x)$. Then, the cumulative density function (cdf) of both X and Y will be $F_X(x)$ and $F_Y(y)$, respectively. It is easy to show that for the transformation h to produce the distribution $F_Y(y)$, starting from the uniform distribution $F_X(x)$, one must have [47]:

$$y = F_Y^{-1}(x) . \tag{3.51}$$

As an example, consider the Laplace distribution that will be obtained from the delta potential well version of the quantum PSO algorithm in Ch. 5. The pdf of the particle trajectory can be written as:

$$f\left(r_k^{i,n} \middle| r_{k-1}^{i,n}\right) = \frac{g \ln \sqrt{2}}{\left|r_{k-1}^{i,n}\right|} \exp\left\{-2g \ln \sqrt{2} \frac{\left|r_k^{i,n}\right|}{\left|r_{k-1}^{i,n}\right|}\right\} , \tag{3.52}$$

where $r_k^{i,n}$ is defined as in (3.46) and g is the only control parameter in the quantum PSO. By solving (3.51), denoting the uniform random variable by u, and the position r by y, one obtains:

$$x_k^{i,n} = \begin{cases} p_{k-1}^{i,n} + \frac{\left|x_{k-1}^{i,n} - p_{k-2}^{i,n}\right|}{2} \ln(2u) & 0 < u \le \frac{1}{2} . \\ p_{k-1}^{i,n} + \frac{\left|x_{k-1}^{i,n} - p_{k-2}^{i,n}\right|}{2} \ln \frac{1}{2(1-u)} & \frac{1}{2} < u \le 1 . \end{cases} \tag{3.53}$$

This alternative derivation of the update equations was accomplished without reference to any argument in physics; the concept of collapsing the wave function, as presented in [42], has not been mentioned at all; only probability theory is needed in arriving to (3.53). Moreover, notice that the control parameter g cancels out in the last expression. The new update equations above were tested numerically and demonstrated a performance as good as the original quantum PSO.

3.7.3 Markov Chain Model

A Markov process is defined as a stochastic process whose past has no influence on the future if its present is specified [47]. Mathematically, we state this definition as follows:

$$P\{\boldsymbol{x}_n \le x_n | \boldsymbol{x}_{n-1}, \boldsymbol{x}_{n-2}, ..., \boldsymbol{x}_1, \boldsymbol{x}_0\} = P\{\boldsymbol{x}_n \le x_n | \boldsymbol{x}_{n-1}\} , \tag{3.54}$$

where $\boldsymbol{x}_k$, $k = 1, 2, ..., n$, is a sequence of random variables defined over a time index k and P is the probability operator. It is understood that the sequence starts at the initial time $k = 0$, which can be considered the *cause*, and then "propagates" the *effect* through certain stochastic rules. Equation (3.54) tells us that the effect at a certain time instance k is dependent only on the information of the previous time step $k - 1$. If we define the information of the sequence at k to be the state at that time instant, then it is possible to describe the Markov chain as a stochastic process in which the system can remember only the previous state.

From the analysis presented at the previous parts of this chapter, we have found that the PSO algorithm, being classic or quantum, involves a conditional pdf in which information about the current position depends only on information coming from the previous time step. In other words, by examining some of the distributions presented here, such as (3.49) and (3.52), it is clear that only the previous state (iteration) will be involved in the evaluation of the probability of the next iteration. Therefore, we classify all of the possible versions of the PSO algorithm, being classic or quantum, as stochastic Markov chains.

It is clear that the functional form of the conditional pdf does depend on the time index k. This is because the new center of gravity $p_k^{i,n}$ [see Eq. (3.46)] varies from iteration to iteration. A Markov process that satisfies such a property is called inhomogeneous Markov process [47].

From the chain rule in probability theory, one can write [47]:

$$f(x_k, x_{k-1}, \ldots, x_1, x_0) = f(x_k | x_{k-1}) f(x_{k-1} | x_{k-2}) \cdots f(x_2 | x_1) f(x_0) , \qquad (3.55)$$

where f stands for the pdf of the random variables presented at its argument. Equation (3.55) implies that the joint statistics of all the positions, evaluated at all iterations up to the present time, can be expressed in terms of the conditional pdf, at various previous times, plus the initial distribution at the first iteration $f(\mathbf{x}_0)$. This initial condition is independent of the algorithm and must be chosen by the user. It is customary in evolutionary methods to select the starting population as a uniform random distribution over the range of interest.

Of special importance in Markov chains is the following result, known as Chapman-Kolmogorov equation [47]:

$$f(x_n | \mathbf{x}_k) = \int_{-\infty}^{\infty} d^N x_m \, f(\mathbf{x}_n | \mathbf{x}_m) f(\mathbf{x}_m | \mathbf{x}_k) , \qquad (3.56)$$

where $n > m > k$. Here $d^N x_m$ stands for the volume element (scalar) in the abstract N-dimensional space $\mathbf{R}^N$. Equation (3.56) can be used to establish a *causal* link between any combination of three iterations that need not to be successive.

For the general case of k successive steps, it is easy to calculate the marginal pdf of the kth state as follows:

$$f(\mathbf{x}_k) = \int_{x_{k-1}} \int_{x_{k-2}} \cdots \int_{x_0} d^N x_{k-1} \, d^N x_{k-2} \ldots d^N x_0 \, f(\mathbf{x}_k, \mathbf{x}_{k-1}, \ldots, \mathbf{x}_1, \mathbf{x}_0) . \qquad (3.57)$$

By substituting (3.56) into (3.57) we get:

$$f(\mathbf{x}_k) = \int_{x_{k-1}} \int_{x_{k-2}} \cdots \int_{x_0} d^N x_{k-1} \, d^N x_{k-2} \ldots d^N x_2 \, d^N x_0 \, f(\mathbf{x}_k | \mathbf{x}_{k-1})$$
$$\cdots f(\mathbf{x}_{k-1} | \mathbf{x}_{k-2}) f(\mathbf{x}_2 | \mathbf{x}_1) f(\mathbf{x}_0) . \qquad (3.58)$$

This gives the pdf of the future kth iteration by causally accumulating the effects of all previous states (iterations) as originated from an initial distribution.

3.7.4 Generalized PSO Algorithm

It is clear from Eq. (3.58) that knowledge of the *conditional* pdf of the sequence of positions, together with the pdf of the initial distribution, is enough to give an exact characterization of the total pdf for any future iteration. Based on these facts, we propose a generalized PSO algorithm summarized as follows.

1. Specify an initial distribution with known pdf $f(\mathbf{x}_0)$.

2. Derive the conditional pdf $f(\mathbf{x}_k | \mathbf{x}_{k-1})$, for arbitrary k, by solving the diffusion equation (classic PSO) or Schrödinger's equation (quantum PSO).

3. Calculate the marginal pdf $f(\mathbf{x}_k)$ using Eq. (3.58).

4. Update the position by realizing the random number generator with the pdf $f(\mathbf{x}_k)$ (for example, the method presented in Sec. 3.7.2).

The generalized PSO algorithm, as presented above, has the following advantages over the traditional classical or quantum versions.

1. The velocity terms are canceled. If the derived update equations of the position are not complicated, then this may considerably reduce the computational complexity of the new algorithm.

2. The algorithm is purely stochastic. The rich literature on stability and time evolution of stochastic processes can be applied to provide a deeper understanding of the convergence of the PSO algorithm.

3. The generalized PSO algorithm eliminates the variations in the physics behind the method; it shows that all PSO methods are special classes within the wider group of Markov chains.

3.7.5 Prospectus and Some Speculations

In particular, it can be inferred from the overall discussion of this section that the Markov approach unifies the classical and the quantum versions of the swarm mechanics, a problem that is fundamental in theoretical physics. It is interesting to refer to the work proposed, yet in a different context, by Nelson in 1966 [54]. The idea was to derive quantum mechanics from a classical model starting from the stochastic diffusion equation. Even though not all physicists have accepted the idea, very recently a general approach based on stochastic dynamics was proposed to provide a theoretical derivation of quantum mechanics by formulating the problem as a stochastic dynamical system obeying some reasonable energy-related principles [55]. If quantum mechanics can be derived from the diffusion equation, then the diffusion theory of Sec. 3.6 and the Markov model of this section are enough to theoretically derive any possible PSO algorithm that can be mapped to existing physical theory. A general approach to establish physical theory for natural selection was formulated by Lee Smolin [56] where it can be seen that evolution, as described in biology, could be linked, on the fundamental level, to the Standard Model of particle physics. Similar ideas, besides the approach presented in our work, are in conformity with each other and strongly suggest that the foundations of the swarm intelligence methods could be rooted ultimately in physics, rather than biology.

CHAPTER 4

Boundary Conditions for the Particle Swarm Optimization Method

4.1 INTRODUCTION

The PSO algorithm belongs to a group of methods in engineering called meta-heuristics. Although such generic techniques are based on several approximations, they still work efficiently with complex problems encountered in various practical setups. However, the drawback of the relative conceptual simplicity of such methods is the emergence of many control parameters that need to be tuned by hand. Even though the PSO has less number of parameters compared with the genetic algorithm (GA), it still requires considerable experience to learn how to tune it for varying applications. Among the many parameters that should be specified by the user is the way the boundaries of the search space are handled. The reason is that the final obtained solution must lie within the allowable limits dictated by the physics of the problem under consideration.

The traditional velocity-clipping criterion, which will be described later (see Fig. 4.2), is the common boundary condition adopted by most researchers and users of the PSO method [3]. Recently, it was demonstrated that the choice of the correct boundary condition can be very critical [57]. Although there are different ways in which these boundaries are handled in literature, few investigations are available on how different boundary conditions perform in engineering problems [7], [70], [27].

For example, Fig. 4.1 illustrates two classes of boundary condition, the *restricted* and and the *unrestricted* boundary modes [27]. The idea for the restricted boundary condition is the following. Once a particle hits the boundary of the search space, it will be re-inserted into the allowable domain by some treatment of the velocity vector (e.g., absorbing, reflecting, etc.). Therefore, the particle stays in the required physical search range and a fitness function evaluation is assigned to it in every iteration. The situation in the unrestricted boundary mode is slightly different. Once a particle hits the boundary walls, no fitness function is evaluated for that specific particle. This prevents "bad" particles from participating in the global interaction of the swarm for the subsequent periods of the trajectory history. This unrestricted mode is certainly more attractive for electromagnetic applications since the highest cost in the optimization process resides precisely in the fitness function evaluation.

One of the serious problems in any optimization method is how to define in the algorithm the allowable solution space. In other words, if the problem is N-dimensional, then the optimization method should try solutions that belong only to a specified subspace D, such that $D \subset \mathbf{R}^N$. The

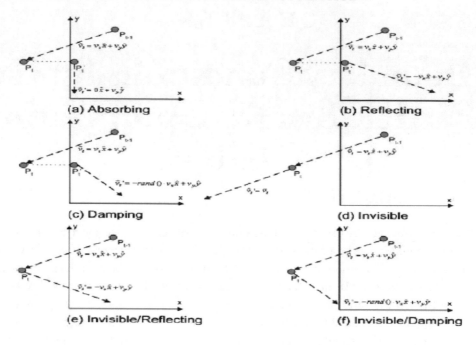

Figure 4.1: Various boundary condition studied in [27]. (Reprinted here with permission from *IEEE Trans. Antennas Propagat.*, which is in [27].)

$$\text{if } v_n^m \geq V_{\max} \text{ then } v_n^m = V_{\max}$$
$$\text{else if } v_n^m \leq -V_{\max} \text{ then } v_n^m = -V_{\max}$$
$$\text{end if}$$

Figure 4.2: Velocity clipping to prevent explosion.

definition of D depends on the physics of the problem in hand. This problem becomes more serious with evolutionary computing techniques since solution candidates must be allowed to move inside a finite domain. The larger this domain, the more difficult, and consequently more expensive, becomes the search process.

Traditionally, in the PSO method particles are initialized randomly in a region supplied by the user. It is assumed, therefore, that the algorithm should find the global optimum inside this region. However, nothing can prevent particles from going outside and converge to an optimum that does not satisfy the set of constraints imposed by the physical problem.

$$\text{if } \left(x_n^m \geq x_n^{\max} \text{ OR } x_n^m \leq x_n^{\min} \right)$$

$$\text{then } v_n^m = -v_n^m$$

$$\text{end if}$$

(a)

$$\text{if } \left(x_n^m \geq x_n^{\max} \text{ OR } x_n^m \leq x_n^{\min} \right)$$

$$\text{then } v_n^m = 0$$

$$\text{end if}$$

(b)

Figure 4.3: (a) Reflection boundary conditions (RBC); (b) absorbing boundary condition (ABC).

In this chapter, we present techniques to truncate the domain of interest and propose a terminology for general boundary conditions that can successfully confine the particles. Following the main theme introduced formally in Ch. 3, we build on certain analogies with physical systems to devise new insights and methods for dealing with the boundary condition problem in the PSO.

4.2   THE SOFT CONDITIONS

The method widely used in PSO literature to prevent particles from exploding outside the physical domain is the introduction of maximum velocity in each dimension, as shown in Fig. 4.2. This means that each particle is not allowed to have velocities that exceed in magnitude a user-defined value $V_{\max}$. Experimental studies of the algorithm show that most of the time particles are confined in the domain of interest. In the majority of the applications tried before, researchers have set $V_{\max}$ to the maximum value of the corresponding coordinate [7].

One improvement suggested recently to enhance the performance of the velocity-clipping criterion in Fig. 4.2 is to add boundary conditions in the domain of interest [7]. For example, the reflection boundary condition (RBC) and the absorbing boundary condition (ABC), depicted in Fig. 4.3, can confine more particles inside the computational domain. However, none of these boundary conditions can ensure that *all* the particles are inside the physical domain all of the time. To illustrate this point, we consider the trajectories of two particles obtained from the application of PSO to the sphere function (3.34). This function has been chosen because it is continuous, convex and unimodal. Since the minimum is located at the origin, then if the PSO algorithm is initialized in a sub-region $D \subset \mathbf{R}^N$, with the origin in the center of this region, then we expect that most of the time the particles will be attracted, through the mechanism of the swarm intelligence, into the global optimum at this center. Stated differently, because there are no other local or global optima, only random motion can push particles outside their initial domain since there exist no other attractive fields outside this domain.

In this chapter, we define *soft* boundary conditions as the case when particles can exist outside the domain walls (*soft* domain). It is clear then that a *hard* boundary condition is defined using a position-clipping criterion, where each coordinate is clipped if exceed pre-specified limit.

Figure 4.4 illustrates the results of the PSO algorithm applied to the sphere function defined above. Here we show an overlay plot of 10 trajectories obtained by the PSO run with 20 particles, 500 iterations, and with $N = 30$. It is clear from Figs. 4.4(a)–4.4(c) that the soft boundary conditions, even with RBC or ABC, cannot confine the particles to be inside the domain of interest (dashed lines). Figure 4.4(d) shows that a hard domain, together with RBC, does limit the particles to be inside the domain all the time.

Therefore, soft domains, even if equipped with RBC or ABC, cannot prevent all particles from existing outside the domain of interest. The explanation suggested here is that certain time is required for the particle to *decelerate* from previous state to new—opposite—state. In other words, particles in classical Newtonian world cannot change their states in zero time because of their inertia. Therefore, some particles will eventually exist outside the domain walls. In contrast, this will not happen with the hard domain. Therefore, we suggest integrating velocity strategies, like RBC or ABC, with hard wall boundary conditions to help particles that are clipped (stuck) at the domain wall to go back into the computational domain.

4.3   THE HARD BOUNDARY CONDITIONS

The hard boundary condition is any boundary that uses a position-clipping criterion as defined in Fig. 4.5. We will investigate the integration of hard boundary conditions with the reflection (RBC) and absorbing (ABC) boundary conditions. Although RBC and ABC have been applied to benchmark functions [7] but no detailed study has been reported on their application to real problems. By combining the hard wall with these conditions one can arrive to an improved version of the PSO algorithm.

4.4   COMPARATIVE STUDY OF HARD AND SOFT BOUNDARY CONDITIONS

In this section we apply the improved PSO to linear array synthesis problems with uniform separation of 0.5λ. In most applications, researchers have set V_{max} to the maximum value of the corresponding coordinates [9]. This has been found to be suitable for solving large number of problems. However, in this application we suggest varying V_{max} to search for a better solution.

On other hand, when the PSO was run using only the velocity-clipping criterion of Fig. 4.2, the authors noticed that the algorithm quickly stuck at local minima. However, the incorporation of the RBC dramatically improved the performance, as can be inferred from Fig. 4.6(a). In

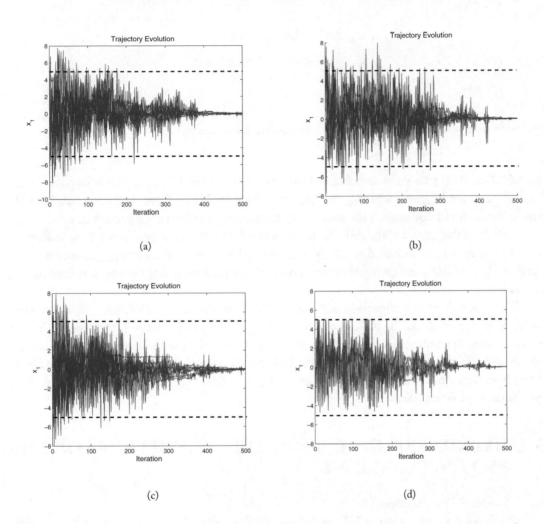

Figure 4.4: Overlay plot of 10 realizations trajectories of the first particle in an optimization of sphere function using PSO algorithm. The dashed line represents the domain wall. The number of dimensions is 30 and each variable is randomly initialized in the range [-5,5]. The PSO uses twenty particles with 500 iterations. The cognitive and social factors are $c_1 = c_2 = 2.0$. The inertia weight w is linearly changed from 0.9 to 0.4. V_{max} is set to the maximum dimension 5. (a) Soft domain without RBC or ABC; (b) soft domain with RBC; (c) soft domain with ABC; (d) hard domain with RBC. (Reprinted here with permission from *Microwave Optic. Technol. Lett.*, which is [57].)

$$\text{if } x_n^m \geq x_n^{\max} \text{ then } x_n^m = x_n^{\max}$$
$$\text{else if } x_n^m \leq x_n^{\min} \text{ then } x_n^m = x_n^{\min}$$
$$\text{end if}$$

Figure 4.5: The position (coordinate) clipping criterion (hard domain).

Figs. 4.6(b)and 4.6(c) we show the obtained pattern, where a deep triangular dip is imposed in the sidelobe region, for $V_{\max} = 1$ and $V_{\max} = 0.3$, respectively. It obvious from the figures that the PSO in the first case could not achieve the same performance compared with the second case.

A study of the impact of the ABC is shown in Fig. 4.7 for the same problem setting in Fig. 4.6. Although the performance stayed relatively unchanging for several values of $V_{\max}$, it suddenly improved as $V_{\max} = 0.05$, indicating that the choice of this parameter can be very sensitive for the problem in hand.

Figure 4.8 shows a comparison between the ABC, RBC, and also with the situation when no other boundary is used. The RBC shows better performance compared with the other technique used here. Also, the author noticed that changing $V_{\max}$ when only the velocity-clipping criterion is used has little effect on the performance, except when gets close to 0.1. Thus, the incorporation of velocity boundaries like RBC or ABC adds extra degree of flexibility in which the parameter $V_{\max}$ plays a more significant role.

4.5   HYBRID PERIODIC BOUNDARY CONDITION FOR THE PSO ENVIRONMENT

4.5.1   General Formulation

In physics, there is a need to apply MD techniques for the study of crystal or liquid bulk materials, where the number of atoms can easily reach 10^{23}. With all possible simplifications in the MD itself, for example ignoring many-body interactions, still a direct simulation of such huge number of particles is impossible, even with the modern increase in available computing resources.

To solve this problem, a physicist conducted the MD study based on one cell only, called the *primary cell*. The idea is to exploit the fact that for crystal-like structures there is a strong spatial pattern repetition or effective periodicity. The primary cell is surrounded from all sides by infinite number of copies, called *image cells*. Therefore, the MD algorithm can update only particles in the primary cell.

Here, we will assume that the PSO particles are moving in a crystal-like structure similar to the one found in natural materials and described above. This is not a necessary restriction in the original PSO. However, we make use of it to suggest a new BC for the basic PSO method.

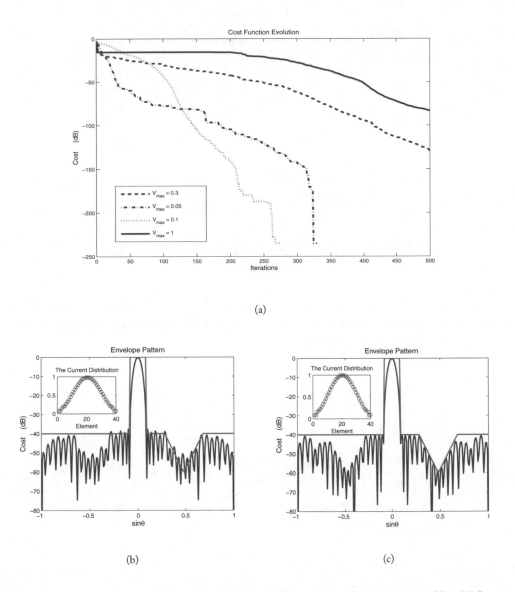

(a)

(b)   (c)

Figure 4.6: Application of the PSO to amplitude-only linear array antenna synthesis. The PSO uses sixty particles with 500 iterations. The cognitive and social factors are $c_1 = c_2 = 2.0$ and the inertia weight w is linearly changed from 0.9 to 0.4. The beamwidth of the pattern is 9° synthesized using 40 elements with enforced symmetry in the current distribution. (a) Averaged (5 realizations) convergence curves obtained using hard walls and RBC with varying V_{max}; (b) obtained pattern for $V_{max} = 1$; (c) obtained pattern for $V_{max} = 0.3$. (Reprinted here with permission from *Microwave Optic. Technol. Lett.*, which is in [57].)

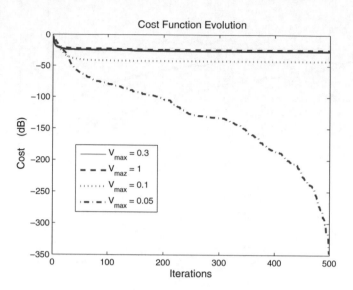

Figure 4.7: Averaged (5 realizations) convergence curves obtained using hard walls and ABC with varying V_{max}. The problem settings match those in Fig. 4.4. (Reprinted here with permission from *Microwave Optic. Technol. Lett.*, which is in [57].)

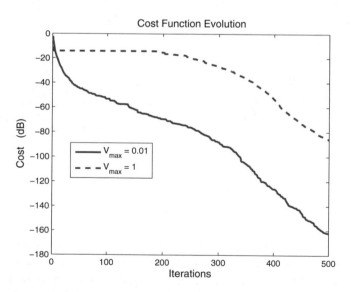

Figure 4.8: Averaged (5 realizations) convergence curves obtained using hard walls and ABC with varying V_{max}. The problem settings match those in Fig. 4.4. (Reprinted here with permission from *Microwave Optic. Technol. Lett.*, which is in [57].)

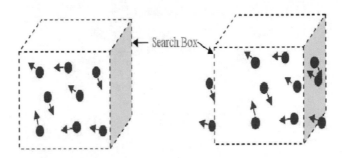

Figure 4.9: The particles in the PSO must be confined within a specific volume dictated by the physical requirements of the optimization problem under consideration. Therefore, an efficient treatment for how particles interact with the domain's walls is critical. (a) Hard boundary conditions (no particle is allowed to leave the search box); (b) soft boundary condition (particles can be allowed to go through the boundary but eventually they should be pulled back). (Reprinted here with permission from *IEEE Trans. Antennas Propagat.*, which is in [69].)

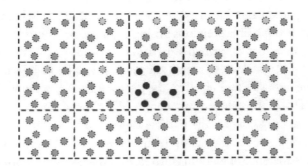

Figure 4.10: Pattern repetition of the proposed periodic PSO environment. The rectangular cell at the middle of the structure, which contains solid particles, is the primary basic PSO cell. The remaining neighboring cells (dashed particles) are the image cells. (Reprinted here with permission from *IEEE Trans. Antennas Propagat.*, which is in [69].)

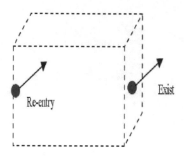

Figure 4.11: A graphical depiction of the periodic boundary condition (PBC). Any particle that leaves the primary cell at one face will be re-inserted at the opposite face. (Reprinted here with permission from *IEEE Trans. Antennas Propagat.*, which is in [69].)

Figure 4.10 illustrates the idea in 2-D space. The primary cell at the middle represents a search box of the PSO defined by the user in accordance to the physical requirements of the problem.

Figure 4.11 illustrates the way in which the PBC is implemented in MD. If the particle leaves the search box from a specific face, then it will be re-inserted from the *opposite* face with the same relative position and velocity. This will insure that the total number of particles will be the same during the simulation. In particular, total energy is conserved within this scheme.

There are, however, two possible scenarios arising in the actual implementation of the above BC. Figure 4.12(a) depicts the first case when a particle falls inside one of the image cells adjacent to the primary cell in a 1-D crystal. In this case, the particle will be re-injected into the primary cell at a distance from the opposite face equal to the distance from the original face in which the particle passed through. This is the case encountered in MD simulations since the time steps used there are very small; it is not possible for the particle to jump far away from its previous position during one time step. However, due to the relatively smaller number of iterations in PSO compared to MD, and also due to the existence of random generators in the particle's motion, it is possible that during a single iteration the particle will jump many cells away from the primary cell as illustrated in Fig. 4.12(b). If the period of the crystal-like structure is P, the particle may fall at a distance $(K + L)P$, where K is an integer and $0 \leq L < 1$. In this case, an image-particle will be inserted at the distance L from the opposite face.

The logic behind this choice is the following. Assume that we gradually decrease the time step. According to the proposed PBC, the particle that passes through one face is re-inserted from the opposite direction. It is possible to imagine that as the particle hit the boundary of the adjacent image cell, its image will hit the original face in the primary cell. Therefore, the situation of a particle

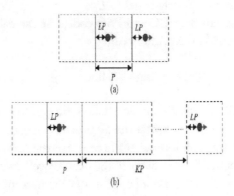

Figure 4.12: 2-D depiction of implementation of the PBC in a 1-D crystal. (a) Illustration for the case when a particle falls outside the primary cell but into one of the adjacent image cells; (b) the case when the particle falls away from the boundary but not at one of the adjacent image cells. (Reprinted here with permission from *IEEE Trans. Antennas Propagat.*, which is in [69].)

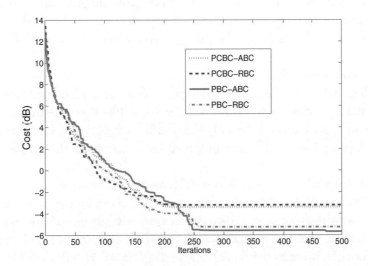

Figure 4.13: Comparison between different hybrid boundary conditions for the PSO working on the 2-dimensional Rastigrin problems. The population size is five particles. The inertia factor w was decreased linearly from 0.9 to 0.2. $c_1 = c_2 = 2.0$. The results are averaged over 100 runs. (Reprinted here with permission from *IEEE Trans. Antennas Propagat.*, which is in [69].)

falling exactly KP away from one face in the primary cell is equivalent to a net movement of zero for the image-particle. Effectively, this means that the image of the particle outside the primary cell in Fig. 4.12(b) will be at a distance exactly equal to L.

To automatically implement the PBC in a PSO code, one can calculate the following *effective distance* of a particle falling outside the primary cell:

$$L = \frac{x}{P} - \text{Sign}(x)\frac{1}{2} - \text{Sign}(x)\,\text{floor}\left(\left|\frac{x}{P} - \text{Sign}(x)\frac{1}{2}\right|\right), \tag{4.1}$$

where floor(r) is a function that returns the smallest nearest integer to r. The function Sign $(\cdot)$ returns 1 when its argument is positive and -1 when the argument is negative. Here, x is the position of the particle assuming that the origin is located at the center of the primary cell. The factor $1/2$ takes into consideration that the origin of x is midway in the primary cell. For example, when x is positive, the difference between $x/P - 1/2$ and floor $(x/P - 1/2)$ is nothing but a compact way to write the desired fraction of the distance L.

In the above formulation of the PBC, a critical assumption has been implied. This is that the force acting on each particle has a limited range. In other words, each particle in the primary cell should in principle be affected by interactions with other particles in the image cell. However, to simplify the PBC, we assume that the force in (3.9) has a limited range; we define a cutoff wall beyond which no interactions are taken into consideration. Similar assumptions have been already made in MD, which resulted in good results for many practical configurations [43]. In this boundary condition, we assume that no particle outside the primary cell can affect the motion of particles inside the primary cell. In spite of its simplicity, the PBC proved to be efficient as would be illustrated later.

4.5.2 Hybrid Implementation

As suggested in Sec. 4.4, if both hard and soft types of BCs are used simultaneously, considerable improvement in the algorithm performance is observed. While there are many different ways to implement SBCs in literature [7], [70], only the position clipping boundary condition (PCBC), shown in Fig. 4.5, is used as a HBC. The proposed PBC represents, then, a new addition to the HBC family.

Based on the authors' experience with several standard test functions and practical engineering problems, it has been found that the performance of the PSO can be greatly enhanced if the HBC and the SBC are used simultaneously. That is, when a particle hits one of the search space domain walls, the algorithm will switch to special routines to handle both the position and the velocity. In this chapter, the format of the new hybrid boundary condition will be 'HBC-SBC.' For example, the single label PBC-ABC denotes a combination of PBC with ABC, where the first boundary refers to hard type while the second to soft type.

4.5.3 Results

We consider a swarm of only five particles applied to solve the two problems defined above for $N = 2$. The reason of choosing small population size is to test how the PBC is capable of modeling a fictitious infinite search space. The search space for the two examples is the interval $[-1, 1]$. Notice that the minimum for the Rastigrin problem is located at the origin, while for the Rosenbrock valley it is at $(1, 1)$. Thus, this search range will test two types of initialization, symmetric and asymmetric modes. This is of special importance to the topic under consideration since the performance of the BC is problem-dependent; i.e., the performance of the PSO, integrated with a specific BC, depends on the location of the optimum with respect to the boundary. It is also a sensitive function of the population size.

In all of the numerical experiments of this section, the social and cognitive parameters are chosen to be $c_1 = c_2 = 2.0$ and the inertia weight w is decreased linearly from 0.9 to 0.2. All results reported are the average of 100 independent runs of the PSO algorithm. Figure 4.13 illustrates a comparison between different boundary conditions for the PSO solution of Rastigrin problem. It is clear that the hybrid combination of the PBC with the ABC or RBC provided better results compared to the PCBC hybrid combinations. Notice that even with two dimensions, the landscape of the fitness functions is highly multimodal. With the small number of particles used, there is a low probability that some particles will explore regions of the search space located far enough from the strong local attractor. However, it seems that with PBC the inherent simulation of infinite environment was able to guide the particles to unexplored regions, resulting in better solutions.

Figure 4.14 illustrates a different experiment.

The minimum of the Rosenbrock valley is located at $(1,1)$. Here, the optimum is closer to one of the search space walls. Therefore, the experiment will test how the algorithm is robust with varying topological conditions in the fitness landscape. The results show that the hybrid combination, in which the PBC is the "hard" one, outperforms considerably other combinations. In these two examples, the boundary PBC-RBC proved to be better than the PBC-ABC.

Next, we consider an array of 14 elements to synthesize a tapered side lobe mask that decreases linearly from -25 dB to -35 dB. The beamwidth of the array pattern is $9°$. The total number of sampling points is 201. A swarm of only eight particles is considered. Figure 4.15 demonstrates the performance of single HBC or SBC. Figure 4.16 shows the performance of various hybrid BCs for PSO solution of the same problem. Again, the advantage gained by using the hybrid combination with the PBC is very clear. Both of the convergence speed and the final error levels are dramatically improved with the introduction of the hybrid PBC. In this example, the formula PBC-RBC proved to be better than the PBC-ABC. Figure 4.17 demonstrates the achieved radiation pattern for the best case in Fig. 4.16. It is clear that all of the design goals are satisfied.

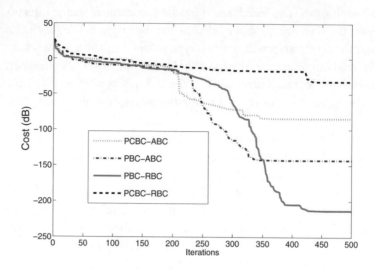

Figure 4.14: Comparison between different hybrid boundary conditions for the PSO working on the 2-dimensional Rosenbrock problem. The population size is five particles. The inertia factor w was decreased linearly from 0.9 to 0.2. $c_1 = c_2 = 2.0$. The results are averaged over 100 runs. (Reprinted here with permission from *IEEE Trans. Antennas Propagat.*, which is in [69].)

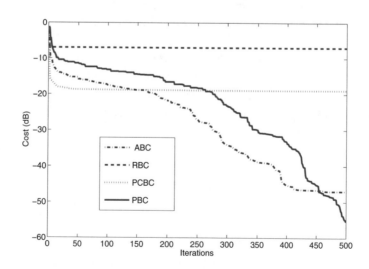

Figure 4.15: Comparison between different single boundary conditions for the PSO working on the problem of the synthesis of symmetric pattern by using a linear array of 14 elements. The population size is eight particles. The inertia factor w was decreased linearly from 0.9 to 0.2. $c_1 = c_2 = 2.0$. (Reprinted here with permission from *IEEE Trans. Antennas Propagat.*, which is in [69].)

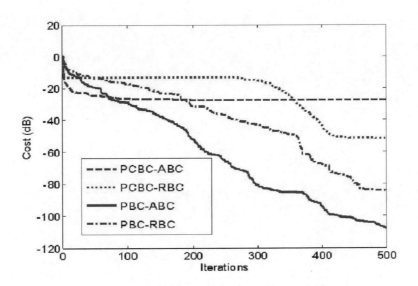

Figure 4.16: Comparison between different hybrid boundary conditions for the PSO working on the problem of the synthesis of symmetric pattern by using a linear array of 14 elements. The population size is eight particles. The inertia factor w was decreased linearly from 0.9 to 0.2. $c_1 = c_2 = 2.0$. (Reprinted here with permission from *IEEE Trans. Antennas Propagat.*, which is in [69].)

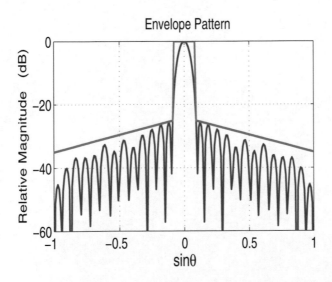

Figure 4.17: The radiation pattern of the linear array antenna corresponding to the PBC-RBC case of Fig. 4.15. (Reprinted here with permission from *IEEE Trans. Antennas Propagat.*, which is in [69].)

CHAPTER 5

The Quantum Particle Swarm Optimization

A quantum-inspired version of the PSO algorithm (QPSO), proposed recently in [40], [41], permits all particles to have a quantum behavior instead of the classical Newtonian dynamics that were assumed so far in many of the common versions of the PSO. Thus, instead of the Newtonian random walk, some sort of "quantum motion" is imposed in the search process. When the QPSO is tested against a set of benchmarking functions, it demonstrated superior performance as compared to the classical PSO but under the condition of large population sizes [40]. One of the most attractive features of the new algorithm is the reduced number of control parameters. Strictly speaking, there is only one parameter required to be tuned in the QPSO.

In this chapter, a generalized framework that allows the user to derive many versions of the QPSO method is proposed where distinct potential well types are presented. A physical interpretation of the algorithm is presented through the discussion of different possible potential wells. Based on our understanding of the physical roots of the new strategy, we propose guidelines to control the tuning parameter of the algorithm. We first introduce the QPSO by illustrating its application to linear array antenna synthesis problems. By conducting several computer experiments and comparing the performances of the two algorithms, the QPSO was found to outperform the classical PSO. Then, the new algorithm is used to study the application of antenna modeling using a set of infinitesimal dipoles. By formulating the modeling of circular dielectric resonator antenna (DRA) as an optimization problem, the QPSO algorithm was able to find a model of 10 dipoles, predicting accurately both the near and far fields. Finally, the method is employed in finding an equivalent circuit to study the resonance of the antenna.

5.1 QUANTUM FORMULATION OF THE SWARM DYNAMICS

The QPSO algorithm allows all particles to move under quantum-mechanical rules rather than the classical Newtonian random motion. In the classical environment, all bees are flying toward the "optimum" location defined by $\mathbf{P}$ in (2.5). The particles are then attracted to this location through the optimization process. Such attraction leads to the global optimum. We remind the reader that $\mathbf{P}$ is nothing but a random average of the global and local bests of the particles of the swarm. In quantum mechanics, the governing equation is the general time-dependent Schrödinger equation

$$j\hbar \frac{\partial}{\partial t} \Psi(\mathbf{r}, t) = \hat{H}(\mathbf{r})\Psi(\mathbf{r}, t) , \qquad (5.1)$$

where $\hat{H}$ is a time-independent Hamiltonian operator given by

$$\hat{H}(\mathbf{r}) = -\frac{\hbar^2}{2m}\nabla^2 + V(\mathbf{r}) , \tag{5.2}$$

where $\hbar$ is the normalized Planck constant, m is the mass of the particle, and $V(\mathbf{r})$ is the potential energy distribution. In the Schrödinger equation, the unknown is the wave function $\Psi(\mathbf{r}, t)$, which has no direct physical meaning. However, its amplitude squared is a probability measure for the particle's motion. By imposing the following normalization condition we can justify such a measure:

$$\int\int\int |\Psi(\mathbf{r}, t)|^2 \, dv , \tag{5.3}$$

where the integration is performed over the entire space.

In the present version of the QPSO algorithm, we apply an attractive potential field that will eventually pull all particles to the location defined by (2.5) [40]. In quantum mechanics, this implies that the potential field will generate bound states [1].

To simplify the formulation, assume that we have the 1-D problem consisting of one particle moving in the dimension r. Let $r = x - p$, where p is the average best given by (2.5). For particles to converge according to the QPSO algorithm, r should approach zero. Therefore, we need to apply an attractive potential field centered at the zero. In principle, any potential *well* can work but the simplest one is the *delta-well* given by [1], [71]:

$$V(r) = -\gamma \, \delta(r) , \tag{5.4}$$

where γ is a positive number proportional to the "depth" of the potential well. The depth is infinite at the origin and zero elsewhere. Thus, the delta potential well is an idealized realization of an infinitely strong attractive potential field that works at a single location.

Assuming the principle of separation of variables, we separate the time dependence of the wave function from the spatial dependence. Substituting the separated form into (5.1) we get [1]:

$$\Psi(r, t) = \psi(r) \, e^{-jEt/\hbar} , \tag{5.5}$$

where E represents the energy of the particle. The wave envelope $\psi(r)$ can be found by solving the following time-independent Schrödinger equation

$$\left\{ -\frac{\hbar^2}{2m}\nabla^2 + V(r) \right\} \psi = E\psi . \tag{5.6}$$

Thus, the solutions are time-independent. Such wavefunctions are called stationary states. Non-stationary states can be formed by superposition of the eigen-solutions obtained from the time-independent Schrödinger equation. However, in this book we consider pure eigen-states only, and more specifically, *bound states*.

5.2   THE CHOICE OF THE POTENTIAL WELL DISTRIBU-TION

The next critical step in deriving the QPSO algorithm is the choice of a suitable attractive potential field that can guarantee bound states for the particles moving in the quantum environment. In the following paragraphs we list some of the possible choices.

5.2.1   The Delta Potential Well

Here, we apply the potential distribution defined in (5.4). This choice leads to the simplest analytical solution possible for Schrödinger equation. It can be shown that under this potential field we get the expression of the probability density function (pdf) as shown in Tab. 5.1 [1, 71]. $L = \hbar^2/m\gamma$ is a parameter called the *characteristic length* of the potential well. Thus, for a given system of particles of mass m we can directly control the characteristic length by varying the "depth" of the well γ.

5.2.2   The Harmonic Oscillator

Another potential distribution, which is very common in quantum mechanics, is the harmonic oscillator potential well given by:

$$V(r) = \frac{1}{2}kr^2 \, , \tag{5.7}$$

where k is a parameter defining the well "depth" or strength. Again, this problem has the following well-known analytical solution [1]:

$$\psi_n(r) = \left(\frac{\alpha}{2^n n! \pi^{\frac{1}{2}}} \right)^{\frac{1}{2}} H_n(\alpha r) e^{-\frac{1}{2}\alpha^2 r^2} \, , \tag{5.8}$$

where $\alpha = \left(mk/\hbar^2 \right)^{1/4}$ and H_n is the Hermite polynomial. Equation (5.8) shows that multiple possible eigen-states exits in this system, each with integer index n. However, we may simplify the problem considerably by assuming that only the lowest possible mode (the ground state $n = 0$) is available. In this case, the Gaussian probability distribution can be obtained as illustrated in Tab. 5.1. Again, the characteristic length of this well can be seen to be $\sqrt{\pi}/\alpha$, a quantity that is directly controlled by the strength of the well k.

5.2.3   The Square Potential Well

A simple nonconfining potential distribution in quantum mechanics is the square well [1] defined:

$$V(r) = \begin{cases} 0, & |r| \leq W/2 \, . \\ V_1, & \text{else} \, . \end{cases} \tag{5.9}$$

It represents energy walls that are used to confine all particles with energy less than V_1 inside the walls located between the points $r = \pm W/2$. Although this potential well is more physically realizable, if compared with the "pathological" delta function well, the solution of Schrödinger equation becomes much more demanding. The state function can be obtained in analytical form, but many modes will be available (excited), depending on the relation between the depth of the well V_1 and its width W [71]. For simplicity, we allow the existence of one mode only, which is the first or the ground mode (even mode) given by the expression shown in Tab. 5.1. Here, a, b, ζ, and η are constants that depend on the choice of γ in the relation $V_1 W = \gamma$. Note that by fixing the product of the depth V_1 and the width W to a constant value γ, we force one mode only to exist [71]. Thus, only the parameter W is needed in the description of the state of the particle.

5.2.4   Other Potential Well Distributions

In quantum mechanics, other attractive potential field distributions that can be used in the implementation of the QPSO are possible. For example, we have the Yukawa potential $V_{\text{Yukawa}}(r) = - V_0 e^{-\mu r}/(\mu r)$; the Gaussian potential $V_{\text{Gaussian}}(r) = - V_0 e^{-(\beta r)^2}$; and the Woods-Saxon potential $V_{\text{W}-\text{S}}(r) = - V_0 \{1 + \exp[(r - R)/a]\}^{-1}$ [1]. However, none of these potentials leads to a tractable analytical solution. As we will see in the following sections, for a derivation of a simple and efficient QPSO algorithm it is crucial to have analytical forms for the state function. This is because such states need to be easily inverted as we will see in the coming parts.

Table 5.1: Summary of the pdf and update equations of the delta, harmonic, and square potential wells.

Potential	pdf	Update Equation
Delta Well	$Q(r) = (1/L) e^{-2\|r\|/L}$	$x_{k+1} = p \pm \dfrac{\ln(1/u)}{2g \ln \sqrt{2}} \|x_k - p\|$
Harmonic Oscillator	$Q(r) = (\alpha/\sqrt{\pi}) e^{-\alpha^2 r^2}$	$x_{k+1} = p \pm \dfrac{\ln(1/u)}{2g \ln \sqrt{2}} \|x_k - p\|$
Square Well	$Q(r) = \begin{cases} \frac{a}{W} \cos^2(\frac{\xi}{W}r), \|r\| \leq W/2 \\ \frac{b}{W} e^{-\frac{\eta}{W}r}, r \geq W/2 \\ \frac{b}{W} e^{\frac{\eta}{W}r}, r \leq -W/2 \end{cases}$	$x_{k+1} = p$ $+ \left\{ \frac{0.6574}{\xi g} \cos^{-1}\left[\pm\sqrt{u}\right] \right.$ $\left. \times \|x_k - p\| \right\}$

5.3   THE COLLAPSE OF THE WAVE FUNCTION

So far we have assumed that all particles behave under the influence of Schrödinger equation solved for several potential well distributions. According to Heisenberg principle of uncertainty [1, 71], it is impossible to define simultaneously both the position and the velocity of a particle with arbitrary accurate precision. Thus, in the QPSO we expect that no velocity term will exist in the basic update equations, in contrast to the case with the classical PSO depicted by Eqs. (2.3) and (2.4). Instead, we need to *measure* the location of the particle. This is the fundamental problem in quantum mechanics. Specifically, measuring devices obey Newtonian laws while the particle itself follows the quantum rules. To interface between the two different worlds, one needs to "collapse" the wave function of a moving particle into the localized space of the measurement. This localization process can be easily accomplished through the following Monte Carlo simulation procedure: (1) generate a random variable uniformly distributed in the local space where the measurement is done; (2) equate the uniform distribution to the true probability distribution estimated by the quantum mechanics; and (3) solve for the position r in terms of the random variables assumed.

By generating a random variable u uniformly distributed between 0 and 1, steps (2) and (3) above lead to the update equations illustrated in Tab. 5.1. Notice that the localization of the probability density function for the square well case requires first to realize that the *natural* characteristic length is the width of the square well itself, W.

5.4   SELECTING THE PARAMETERS OF THE ALGORITHM

A fundamental condition of convergence in any QPSO algorithm is given by:

$$\lim_{t \to \infty} L(t) = 0 \,, \tag{5.10}$$

where $L(t)$ is the time-dependent characteristic length. This can be directly inferred from the update equations in Tab. 5.1 where convergence is clearly understood to be the case with $\mathbf{x}^m \to \mathbf{P}^m$ for all particles $m = 1, 2, .., M$. Thus, we need to enforce a time evolving parameter $L(t)$ such that all particles will eventually arrive to the desired location. To guarantee, on the average, that the next particle will converge, we need the value of $|r_{k+1}|$ at iteration k to be closer to zero. A suitable probabilistic translation for this statement is given by

$$\int_{-\infty}^{|r_k|} Q_{k+1}(r) \, dr > 0.75 \,, \tag{5.11}$$

where $Q_{k+1}(r)$ is the probability density function of the particle at the $(k+1)$th iteration. The reader should notice that convergence obtained under condition (5.11) is guaranteed only in the probabilistic sense. Some particle may diverge, but on the average the algorithm should converge. This relation follows from the fact that the integration from $-\infty$ to 0 will produce 0.5, so the "remaining" probability of 0.5 should be divided between the decision that the particle location in the $(k+1)$th iteration will be either to the left or to the right of $|r_k|$. The condition of (5.11) leads to the desired situation where we have a higher probability of approaching the origin.

By solving (5.11) for the delta potential well, we get;

$$L_{k+1} < \frac{1}{\ln \sqrt{2}} |x_k - p| \,. \tag{5.12}$$

Thus, we may choose;

$$L_{k+1} = \frac{1}{g(\ln \sqrt{2})} |x_k - p| \,. \tag{5.13}$$

The condition in (5.12) is automatically satisfied when

$$g > 0 \,. \tag{5.14}$$

This leads to the Quantum Delta PSO (QDPSO).

The derivation of the corresponding formulas for the harmonic oscillator and the square potential wells requires more effort because of the difficulty in carrying out the integration of (5.11). For the harmonic oscillator, by numerically solving an equation involving the error function, we obtained

$$\alpha = g \frac{0.47694}{|x_k - p|} \,. \tag{5.15}$$

This leads to the Quantum Harmonic Oscillator PSO (QOPSO).

For the square well, we obtained the following through a much longer calculations:

$$W = \frac{0.6574}{g} |x_k - p| \,. \tag{5.16}$$

This leads to the Quantum Square-Well PSO (QSPSO). Notice that condition (5.14) is required in (5.15) and (5.16). Table 5.1 summarizes the pdfs and the update equations of the various potential wells derived above.

5.5   THE QPSO ALGORITHM

5.5.1   The Algorithm

Now, it is possible to summarize the proposed QPSO algorithm as follows:

1. Choose a suitable attractive potential well centered around the vector $\mathbf{P}$ given by Eq. (2.5) (Delta well, harmonic oscillator, square well, combinations of the previous wells, etc.). Solve Schrödinger equation to get the wave function, and then the pdf of the position of the particle.

2. Use Monte Carlo simulation—or any other measurement method—to collapse the wave function into the desired region. The result of this step is an equation in the form

$$\mathbf{x} = \mathbf{P} + f(L, \pm \mathbf{u}) \,. \tag{5.17}$$

3. Apply the pseudo-code shown in Fig. 5.1.

Note that the cognitive and social parameters do not appear in the QPSO algorithm. This is clear for the case of $c_1 = c_2$ where from Eq. (2.5) these factors cancel. Most of the research papers on the classical PSO shows that these two factors are better to be chosen equal [7], making this version of the QPSO justified. Values for the cognitive and social coefficients that are not equal to each other were used, but poor performance was obtained for all objective functions chosen. Thus, the generalized QPSO shown above contains one control parameter only, g, which is directly related to the characteristic length of the potential well. This makes the QPSO more attractive for electromagnetic applications compared with the classical PSO that requires extra parameters to be tuned for each application.

5.5.2   Physical Interpretation

Figure 5.2 shows plots for probability density functions for the three potential wells derived above. The functions are normalized, in amplitude and position, to the characteristic length of the well in order to understand the qualitative differences between them. It is clear that the harmonic oscillator leads to the "tightest" distribution around the origin, and thus more particles are likely to be close to p, resulting in faster convergence according to (5.10). However, in a phenomenon similar to the well-known tunneling effect in quantum mechanics [1], some particles with very small but nonvanishing probabilities are allowed to explore regions far away from the origin (the center of the potential well). This is in direct contrast with the rules of classical mechanics where the attractive field will eventually pull all particles that do not have sufficient energies to escape. Thus, we expect that the QPSO will have a much stronger "insight" on the optimization space since it can "spy" on far regions in the domain of interest all the time. The "spies" are very few particles existing in the tail region of the probability density functions.

5.6   APPLICATION OF THE QPSO ALGORITHM TO ARRAY ANTENNA SYNTHESIS PROBLEMS

Both of the classical PSO and the QPSO were tested first against a set of benchmark functions. The functions tried include the sphere function, Rastrigin's function, and Rosenbrock's valley [7], [40]. Other functions were used to examine specific aspects in the performance of the algorithm. Asymmetric initialization ranges were employed to test the origin-bias seeking of the optimization. The algorithm performance in these problems was satisfactory enough to start applying the method to practical problems.

In improving the classical PSO to test its competence with the new QPSO scheme, the same level of complexity in both algorithms is maintained. That is, although there are some advanced supervisory methods suggested in literature to enhance the convergence of the PSO under varying possible physical problems [6], [59], [60], such methods increase the complexity—and thus the computational cost—of the algorithm. Therefore, our comparison is based on maintaining the same level of simplicity in both schemes to guarantee reasonable measures for the performance of the new algorithm compared with the classical version. Also, several runs with varying seeds for the random number generators were reported in the comparison.

5.6.1   General Description of the Linear Array Synthesis Problem

We consider now the application of the QPSO algorithm to linear array antenna design problems. In this part, the consideration focuses on shaping the main beam and the side lobe pattern of the array to meet desired characteristics given by a user-defined function $T(t)$. For further details about the array antenna synthesis problem, see Sec. 3.5.

We will compare the performance of quantum and classical PSO algorithms in achieving certain desired patterns. To establish the comparison on a fair ground, a clipping criterion (hard domain) is employed in both algorithms that will not allow candidate solutions to pass certain

Inialize x^m, $\mathbf{P}^m_{local}$, $\mathbf{P}_{global}$

Do $i = 1, N_{itr}$

   Do $m = 1, N_{population}$

      Update $\mathbf{P}^m_{local}$ and $\mathbf{P}_{global}$

      $\varphi_1 = rand(0,1), \quad \varphi_2 = rand(0,1)$

      $P = \dfrac{\varphi_1 \mathbf{P}^m_{local} + \varphi_2 \mathbf{P}_{global}}{\varphi_1 + \varphi_2}$

      $u = rand(0,1)$

      $L = L\left(g, u, |x^m - p|\right)$

      if $rand(0,1) > 0.5$

         $x = p + f(L, u)$

      else

         $x = p + f(L, -u)$

      end if

   end Do

end Do

Figure 5.1: Pseudo-code for the QPSO algorithm with the explicit form of the functions L and f; the QDPSO, QOPSO, and QSPSO as described by Tab. 5.1. (Reprinted here with permission from *IEEE Antennas Wireless Propagat. Lett.*, which is in [24].)

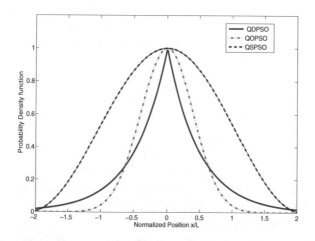

Figure 5.2: Pseudo-code for the QPSO algorithm with the explicit form of the functions L and f; the QDPSO, QOPSO, and QSPSO as described by Tab. 5.1. (Reprinted here with permission from *IEEE Antennas Wireless Propagat. Lett.*, which is in [24].)

boundaries [57]. If a particle hits a specific boundary, its value will be fixed at the value of that boundary, and eventually it should go back to the solution space searching for the global minimum that exists there. All comparisons between different algorithms were reported using the same number of iterations and population size. Preliminary experiments with the QPSO algorithm, when applied to benchmark functions, suggest using high number of population size in order to achieve a satisfactory performance. Based on this, we fix the number of populations to 60 particles in the coming examples. This enhances the "averaging" or the smoothing of the convergence curves obtained. In the following examples, various realizations of the same experiment—starting with different seeds—produced results that are close to each other. An averaging of ten runs is considered and found to be sufficient.

The number of fitness calls equals $N_{\text{iteration}} \times N_{\text{population}}$, where $N_{\text{iteration}}$ is the number of iterations and $N_{\text{population}}$ is the population size. Thus, increasing the population size will dramatically increase the number of fitness calls. Since the calculation of the objective function is the bottleneck in most EM applications, this will make the optimization process increasingly expensive. Therefore, although increasing the number of particles in the swarm will intuitively increase the possibility to reach the global optimum, for practical EM problems the population size should be kept as small as possible.

5.6.2   Optimizing Sidelobe Patterns

The objective function used here is not directly proportional to the absolute difference between the obtained and the desired pattern. Instead, the "Do not Exceed criterion" has been utilized in the formulation of the objective function.

Figures 5.3 and 5.4 illustrate the optimization of 40-element linear array antenna with tapered side-lobe performance. In Fig. 5.3, the obtained array pattern is shown together with the resulted current distribution. In this problem we exploited the symmetry of the current distribution since the desired envelope is already symmetric. This reduces the degrees of freedom into half of the number of array elements.

Both of the classical PSO algorithms, used for the comparison, and the proposed QPSO, employ *asynchronous* velocity update. That is, the update of the global best performance is done after each particle movement, rather than waiting until the required information is collected from the entire particle swarm at the end of the current generation [58]. This technique has been adopted in electromagnetic applications [7], [9] but very recently some researchers [15] reported a detailed study showing that it is more powerful than synchronous update most of the time. Moreover, in the classical PSO version employed here, we integrated what we defined as *hard* domain instead of *soft* domain, leading to an improved classical PSO version [57]. In this version, the value of the maximum velocity V_{max} is changed to enhance the performance. Also, the reflection boundary condition (RBC) is combined with the hard domain. Figure 5.4 clearly shows that when PSO is run without RBC, the algorithm found itself trapped in a local minimum while the QDPSO achieves much better solutions. Thus, by enabling the RBC feature and changing V_{max}, we could obtain much better

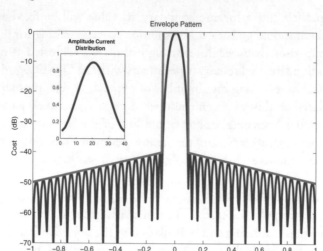

Figure 5.3: Amplitude-only synthesis with tapered side lobes and enforced symmetry. The number of elements is 40 element and the beamwidth is 9.5°. The radiation pattern and current distribution were obtained using QDPSO algorithm with 1,000 iterations, population size of 60, 301 observation points, and characteristic length of $g = 3$. (Reprinted here with permission from *IEEE Antennas Wireless Propagat. Lett.*, which is in [24].)

solution; however, the best solution at $V_{max} = 0.15$ could not compete with the QDPSO, which was obtained using simpler tuning strategy. The only parameter of the algorithm g was tuned by direct trial and error strategy through testing problems with small number of generations. Then, we run the optimization, using the best obtained, for high number of iterations. This is much easier compared with the genetic algorithms (GA) and classical PSO where several tuning parameters should be chosen, each with various possible values and directions of change.

Figures 5.5 and 5.6 illustrate the comparison between PSO and QPSO for more challenging synthesis requirements. Here we impose a wide null in the side-lobe region as shown in Fig. 5.5. The performance of the QDPSO and the classical PSO is reported. The quantum version scores much better for the same number of iterations and population size. In Fig. 5.6, we also show the effect of varying the characteristic length g linearly from 2.5 to 4. It is clear that the QDPSO achieves better convergence rate than the PSO.

The linear variations of g can be justified conceptually if one notices that by such variation the corresponding characteristic length of the well will decrease with iterations. As discussed before, by decreasing the characteristic length, we increase the probability of finding the particle around the origin, thus highlighting the *local* search capabilities of the algorithm. This enhances the performance since with increasing iterations the particles become more close to the global minimum. If one is

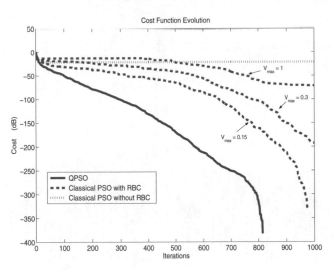

Figure 5.4: Convergence curves for the QDPSO and the classical PSO in the problem of Fig. 5.3. The classical PSO uses the same number of iterations and population size. The inertia weight w is varied linearly from 0.9 to 0.2, $c_1 = c_2 = 2.0$, and the maxim velocity limit V_{max} is changed as indicated. (Reprinted here with permission from *IEEE Antennas Wireless Propagat. Lett.*, which is in [24].)

interested in conducting the comparison only through the first 1,200 iterations, then it seems then that by allowing linear variation of the control parameter the algorithm looks by itself for the best g. The user can later try different values for the minimum and maximum limits of this parameter to narrow the search range, provided that improvement in the performance is possible.

5.6.3 Synthesized Main Beam Patterns
Shaping main beams requires minimizing the *absolute* difference between the obtained array pattern and the desired pattern shape. The cost function is defined as:

$$F = \frac{1}{M} \sum_{k=1}^{M} [|AF(u_k)| - T(u_k)]^2 . \tag{5.18}$$

Using "Do not Exceed criterion" in the side lobe region and the above absolute error criterion in the main beam region leads to multi-objective optimization. In general, this will increase the difficulty of the problem, reducing the chances of reaching to the global minimum. The authors have tried to use the aggregation method by transforming the two cost functions into the following single cost measure $F = \alpha_1 F_1 + \alpha_2 F_2$, where F_1 and F_2 are the cost functions in the main beam and the side lobe regions, respectively. The constants α_1 and α_2 are the possible weights of each objective function, F_1 and F_2, respectively. Optimizing the problem this way did not lead to satisfactory results

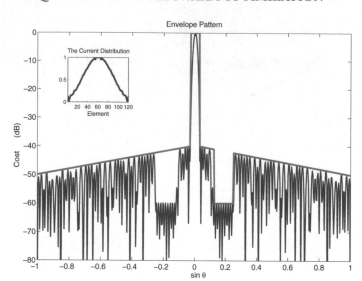

Figure 5.5: Amplitude-only (enforced symmetry) synthesis with tapered side-lobe and deep null. The number of elements is elements with beamwidth of 3°. The radiation pattern and current distribution were obtained using QDPSO algorithm with with 3,000 iterations, population size of 60, 401 observation points, and characteristic length of $g = 3$. (Reprinted here with permission from *IEEE Antennas Wireless Propagat. Lett.*, which is in [24].)

with either the quantum or the classical PSO algorithms. Some prior knowledge about the weights above is needed to formulate the cost function. It is clear that those multi-objective optimization problems cannot lead in general to solutions that satisfy all sub-measures simultaneously. To solve this problem, a single objective function, as defined by (5.18), is used in the entire region. The results are shown in Fig. 5.7 where the goal is to synthesize asymmetrical pattern by using both the amplitude and phase of each array element. Notice that since we are trying to force the array pattern to follow the desired target, there will be certain amount of ripples around this target. This is in contrast to the previous case when the "Do not Exceed criterion" has been used. From Fig. 5.8, we can see that the QDPSO is much faster in convergence compared with the classical version.

In our experiments, the QDPSO appears to be more stable and easy to tune compared with the quantum harmonic oscillator version (QOPSO). The performance of the square well (QSPSO) algorithm was poor for most of the cases tried. The authors noticed that by selecting the width of the square well to be very small, the potential distribution approaches the delta function and the performance improves. However, according the pdf of the QSPSO in Tab. 5.1, the corresponding density function shrinks also by the same ratio, thus increasing the probability of finding more particles around the origin. This leads to very fast convergence but with higher chances to be trapped

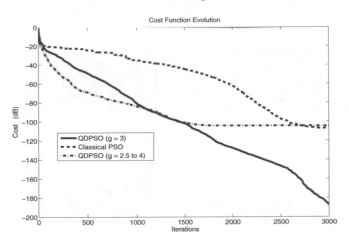

Figure 5.6: Convergence curves for the QDPSO and the classical PSO of the problem of Fig. 5.5. The classical algorithm uses the same number of iterations and population size as the quantum one. The inertia weight w is varied linearly from 0.9 to 0.2, $c_1 = c_2 = 2.0$, and $V_{max} = 0.3$ with RBC. (Reprinted here with permission from *IEEE Antennas Wireless Propagat. Lett.*, which is in [24].)

in local minima since fewer particles are allowed to "spy" on other regions in the solution domain. In some optimization problems, one may be interested in very fast convergence rather than achieving global low cost function (global solution). Thus, the QSPSO can be utilized in such applications.

In spite of the above difficulties with the QOPSO and the QSPSO, it is noticed that these algorithms might compete in other types of applications. Thus, it is up to the user to make the choice of the proper version of QPSO to better suite his problem. Such results are omitted from this chapter for brevity.

5.7 INFINITESIMAL DIPOLES EQUIVALENT TO PRACTICAL ANTENNAS

5.7.1 Formulation of the Problem

To demonstrate the applicability of the new quantum PSO method to different types of applications, we consider in this section the problem of finding a set of infinitesimal dipoles that can model an arbitrary practical antenna configuration, which are realistic antennas with practical gain and radiation patterns. The modeling here means that the obtained set of dipoles must be capable of producing both the near and far fields of the antenna under consideration. This concept was introduced in [61] with both electric and magnetic dipoles used to model the near field of a radiating structure. In [62], the problem was formulated as an optimization problem solved using evolutionary genetic algorithm (GA) techniques. The GA was introduced to tackle a general inverse source

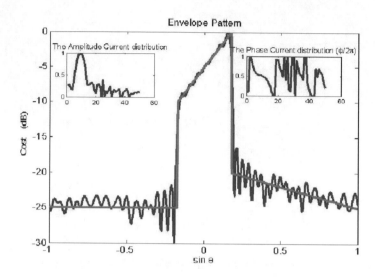

Figure 5.7: The obtained pattern and the current distribution of the complex synthesis of 50-element linear array with beamwidth of 20° using the QDPSO algorithm with 2,000 iterations, population size of 60, 501 observation points, and characteristic length of $g = 3$. (Reprinted here with permission from *IEEE Antennas Wireless Propagat. Lett.,* which is in [24].)

modeling problem using infinitesimal dipoles with both electric and magnetic types [63]. However, the main purpose there was to predict the far field performance of a device under test (DUT) based on the measured near field. In this section, starting from a known near field distribution of an antenna, infinitesimal dipoles are obtained to provide the same near field and consequently the same far field.

Assume that the antenna has the near field distribution $(\mathbf{E}^a, \mathbf{H}^a)$ at a certain surface S. Assume also a set of dipoles $\{\chi_i\}_{i=1}^N$, where N is the number of dipoles and is χ_i a 9-element vector representing the parameters of the ith dipole given by:

$$\chi_i = [\text{Re}\{M_u^i\}, \quad \text{Im}\{M_u^i\}, \quad r^i]^T,$$
(5.19)

where $u = x, y, z$ and $r^i = (x^i, y^i, z^i)$. Here, M_u^i is the ith dipole moments located at the positions u^i, which are constrained by the actual antenna size. The fields generated by the dipoles are denoted by $(\mathbf{E}^d, \mathbf{H}^d)$. The cost function is defined as follows:

$$F = \left[\frac{1}{N_{\text{ops}} \sum\limits_{n=1}^{N_{\text{ops}}} \sum\limits_u \left| E_u^a(\mathbf{r}_n) \right|^2} \sum\limits_{n=1}^{N_{\text{ops}}} \sum\limits_u \left| E_u^a(\mathbf{r}_n) - E_x^d(\mathbf{r}_n) \right|^2 \right]^{1/2},$$
(5.20)

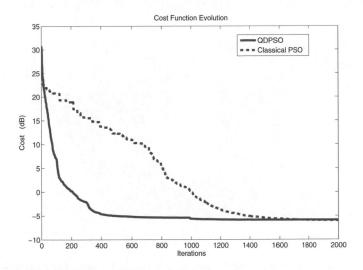

Figure 5.8: Convergence curves for the QDPSO and the classical PSO in the problem described in Fig. 5.7. The classical and quantum algorithm use the same number of iterations and population size. The inertia weight w is varied linearly from 0.9 to 0.2, $c_1 = c_2 = 2.0$, and $V_{max} = 0.3$ with RBC. (Reprinted here with permission from *IEEE Antennas Wireless Propagat. Lett.*, which is in [24].)

where N_{ops} is the number of observation (samples) points, and $\mathbf{r}_n$ is the position vector of the nth sampling point where these set of points are assumed to sample the observation surface S. Notice that the objective function defined in (5.20) uses only the electric field. Once the electric field is synthesized by the obtained set of infinitesimal dipoles, the magnetic field will be satisfied automatically because both fields must satisfy Maxwell's equations.

In general, this cost measure is highly nonlinear, with an objective function landscape full of local minima, making the optimization problem very difficult unless a powerful global search method is used. Once the infinitesimal dipoles are found, both the near and far fields can easily be computed.

5.7.2   Infinitesimal Dipole Model of Circular Dielectric Resonator Antenna

The procedure above is applied to a dielectric resonator antenna (DRA). Figure 5.9 shows the configuration of a circular DRA located above an infinite ground plane. A coaxial probe excites the antenna, as shown in Fig. 5.10. An accurate Method of Moment (MOM) procedure [64] is used to analyze the structure. The DRA is tuned to resonate at the frequency 10 GHz unlike the sample used with the GA [65], which was off-resonance. Near-field data are calculated at a square plane of side length λ. The distance of the observation plane from the ground is taken to be λ, as shown in Fig. 5.10. A set of 10 dipoles is considered. The QDPSO algorithm is used to minimize the

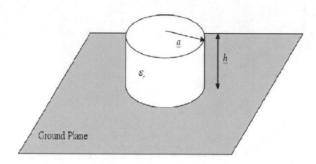

Figure 5.9: Geometry of a circular dielectric resonator antenna located above an infinite ground plane. (Reprinted here with permission from *IEEE Antennas Wireless Propagat. Lett.,* which is in [24].)

objective function defined in (5.20) with the control parameter g set to 3.0 and a population size of eighty particles is considered. The dipoles locations are restricted to be inside the physical domain of the DRA while the dipole moments are chosen based on the order of magnitude of actual near field data.

Figure 5.11 shows the convergence curve obtained for 5,000 iterations. Samples of the obtained near fields are compared with the actual near fields, as shown in Fig. 5.12. Excellent agreement is observed. The RMS error (defined in [65]) computed for this problem is less than 3%. Figure 5.13 shows the comparison between the far-field radiation patterns of the equivalent dipoles and the actual antenna. It should be mentioned that the examples in [65] used only 5 dipoles and the RMS errors was above 7%.

In [63], it was reported that eight dipoles constitute a practical upper limit beyond which convergence problems will start to affect the GA optimization process. However, the new quantum method was able to predict a solution for 10 dipoles using only 1 control parameter, which does not need tuning. The value of the control parameter $g = 3.0$ was used directly based on the previous experience with benchmark functions and the array antenna synthesis problems.

5.7.3   Circuit Model of Circular Dielectric Resonator Antenna

The antenna shown in Figs. 5.9 and 5.10 represents a resonator coupled with the source through a feeding mechanism. Such an antenna can be modeled by lumped-element RLC resonant circuits that resonate at the same frequency of the actual antenna [66, 67].

The probe coupling is considered in the circuit model by adding the series L_c and C_c, as shown in Fig. 5.14. The problem is formulated to minimize the square difference between the input

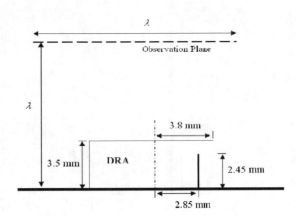

Figure 5.10: Cross-sectional view illustrating the dimensions of the DRA and the location of the monopole deep within the DRA. The dielectric constant of the DRA material is $\epsilon_r = 10.2$. (Reprinted here with permission from *IEEE Antennas Wireless Propagat. Lett.*, which is in [24].)

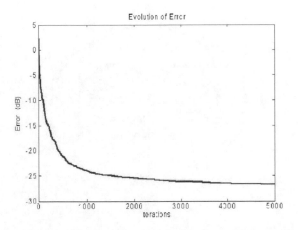

Figure 5.11: Convergence performance of the DRA synthesis problems using a set of 10 dipoles with population size of eighty particles and $g = 3.0$ (Reprinted here with permission from *IEEE Antennas Wireless Propagat. Lett.*, which is in [24].)

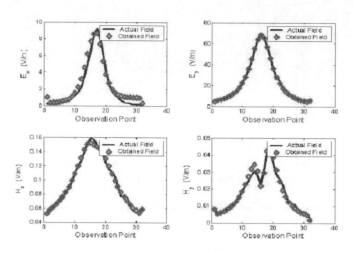

Figure 5.12: Comparison between the actual fields of the DRA (obtained using MOM) and the fields radiated by the obtained infinitesimal dipoles. Near field is computed across a line dividing the square observation plane midway. (Reprinted here with permission from *IEEE Antennas Wireless Propagat. Lett.*, which is in [24].)

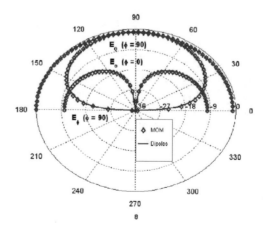

Figure 5.13: Comparison between the actual far fields fields of the DRA (obtained using MOM) and the fields radiated by the obtained infinitesimal dipoles. (Reprinted here with permission from *IEEE Antennas Wireless Propagat. Lett.*, which is in [24].)

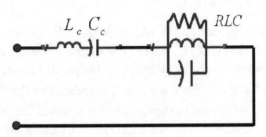

Figure 5.14: Lumped-element circuit model for the DRA including the effect coupling. (Reprinted here with permission from *IEEE Antennas Wireless Propagat. Lett.*, which is in [24].)

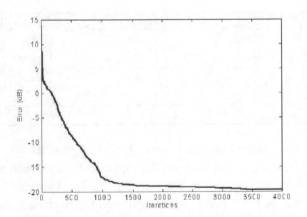

Figure 5.15: Averaged convergence curve for 10 runs of the QDPSO optimization algorithm used to find equivalent circuit model for the DRA in Fig. 5.10. Each run consists of 4,000 iterations in 5-dimensional space. The number of particles is 30. The control parameter is set to $g = 3.0$. (Reprinted here with permission from *IEEE Antennas Wireless Propagat. Lett.*, which is in [24].)

impedance of the equivalent circuit, Z_{in}^e, and the input impedance of the actual antenna, Z_{in}^a, using the following objective function:

$$F = \left[\frac{1}{\frac{N}{\sum\limits_{n=1}} |Z_{in}^a (f_n)|^2} \sum_{n=1}^{N} |Z_{in}^e (f_n) - Z_{in}^a (f_n)|^2 \right]^{1/2} , \tag{5.21}$$

where N is the number of frequency points and f_n is the nth frequency sample. This problem requires optimizing five variables, R, L, C, L_c, c_c. The ranges of these variables must be carefully chosen since they are not independent. For example, for the parallel resonant circuit, the resonance frequency is given by $f_0 = 1/\sqrt{LC}$. Since the actual impedance of the DRA shows resonance at 10 GHz, then both L and C are related to each other. It is assumed that only one mode is excited, which is equivalent to the parallel RLC resonator. Therefore, we considered a small frequency band around the resonance frequency of the antenna. The QDPSO was used to solve the problem with 30 particles, $g = 3.0$, and 4,000 iterations. The range of the parameters and the final values obtained by the optimization are shown in Tab. 5.2.

Table 5.2: The parameters range for each circuit element and the final values obtained after finishing the optimization.

Element	Min Value	Max Value	Obtained Value
R (Ω)	1.00	100.0	56.59
L (nH)	0.01	1.00	0.0724
C (pF)	1.00	10.0	3.4864
L_c (nH)	0.10	10.0	0.5601
C_c (pF)	0.10	10.0	0.3651

Figure 5.15 illustrates the convergence curve QDPSO algorithm with 30 particles, $g = 3.0$, and 4,000 iterations. The comparison between the input impedance obtained using MOM [64] and the one obtained using the equivalent circuit of Fig. 5.14 is presented in Fig. 5.16. It is clear that the predicted input impedance is in excellent agreement with the accurate MOM calculation. Furthermore, by processing the obtained equivalent circuit, it is possible to calculate the unloaded resonance frequency and the radiation quality factor as 10.0164 GHz and 12.4161, respectively. The coupling coefficient and the loaded quality factor are also calculated to be 1.1183 and 5.8613, respectively, [67].

5.8 CONCLUSION

In this chapter, a generalized framework for Quantum Particle Swarm Optimization (QPSO) suitable for electromagnetic problems was proposed. The diversity of the new method was demonstrated

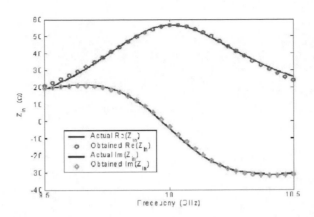

Figure 5.16: Comparison between the input impedance calculated using the circuit obtained by the QDPSO algorithm and the MOM impedance. (Reprinted here with permission from *IEEE Antennas Wireless Propagat. Lett.,* which is in [24].)

by choosing appropriate attractive potential fields, leading to different algorithms were obtained by solving the corresponding Schrödinger equations. The physics of the quantum swarm helped in providing a theoretical explanation of why the algorithm works.

Considering the number of iterations and the population size as common requirements in every evolutionary computing algorithm, we found that the QPSO algorithm contains only one control parameter while its classical counterpart requires the use of four parameters (c_1, c_2, w, and V_{max}) for satisfactory convergence to the desired solution. The computational cost of tuning the four numerical parameters in the classical PSO were very high compared with single-parameter QPSO. In addition, by linearly increasing g, faster convergence could be obtained, but not necessary to global optimum, which could be useful for certain applications when one is looking for sub-optimal solutions. Therefore, the tuning of the QPSO is much simpler and more attractive compared to many of the classical versions of PSO.

The applicability of the new optimization procedure was investigated by studying the problem of synthesis of linear array antennas and the performance of the QPSO was shown to outperform the classical PSO most of the time in the convergence rate as well as the final error level. The QDPSO method was utilized to simulate an actual antenna, such as the circular DRA, by a set of ten infinitesimal dipoles that provide similar near and far fields, which outperformed previous solutions using the GA method applied to the same problem. Also, the QDPSO was utilized to find a circuit model for the DRA. The resonant circuit was used to predict several parameters such as the resonant frequency, quality factors, and coupling coefficient.

Bibliography

[1] F. S. Levin, *An Introduction to Quantum Theory*, Cambridge: Cambridge University Press, 2002.

[2] J. Kennedy and R. C. Eberhart, "Particle swarm optimization," in *Proc. IEEE Conf. Neural Networks* IV, Piscataway, NJ, 1995. 10.1109/ICNN.1995.488968

[3] J. Kennedy and R. C. Eberhart, *Swarm Intelligence*, Morgan Kaufmann Publishers, 2001.

[4] M. Clerc and J. Kennedy, "The particle swarm: explosion, stability, and convergence in a multi-dimensional complex space," *IEEE Trans. Evolution. Computat.*, vol. 6, no. 1, pp. 58–73, Feb. 2002. 10.1109/4235.985692

[5] V. Kadirkamanathan, K. Selvarajah, and P. J. Fleming, "Stability analysis of the particle swarm optimizer," *IEEE Trans. Evolution. Computat.*, vol. 10, no. 3, pp. 245–255, June 2006. 10.1109/TEVC.2005.857077

[6] G. Ciuprina, D. Ioan, and I. Munteanu, "Use of intelligent-particle swarm optimization in electromagnetics," *IEEE Trans. Magn.*, vol. 38, no. 2, pp. 1037–1040, Mar. 2002. 10.1109/20.996266

[7] J. Robinson and Y. Rahmat-Samii, "Particle swarm optimization in electromagnetics," *IEEE Trans. Antennas Propagat.*, vol. 52, pp. 397–407, Feb. 2004. 10.1109/TAP.2004.823969

[8] D. Gies and Y. Rahmat-Samii, "Vector evaluated particle swarm optimization (VEPSO): optimization of radiometer array antenna," *IEEE Int. Symp. Antennas Propagat. Digest*, pp. 2297–2300, 2004.

[9] D. Boeringer and D. Werner, "Particle swarm optimization versus genetic algorithms for phased array synthesis," *IEEE Trans. Antennas Propagat.*, vol. 52, no. 3, pp. 771–779, Mar. 2004. 10.1109/TAP.2004.825102

[10] U. Baumgartner, C. Magele, K. Preis, and W. Renhart, "Particle swarm optimization for pareto optimal solutions in electromagnetic shape design," *IEE Proc. Sci. Meas. Technol.*, vol. 151, no. 6, pp. 499–502, Nov. 2004. 10.1049/ip-smt:20040631

[11] R. Hassan, B. Cohanim, O. de Weck, and G. Venter, "A comparison of particle swarm optimization and the genetic algorithm," *AIAA* 2005 Conference, 2005.

[12] N. Jin and Y. Rahmat-Samii, "Parallel particle swarm optimization and finite-difference time-domain (PSO/FDTD) algorithm for multiband and wide-band patch antenna designs," *IEEE Trans. Antennas Propagat.*, vol. 53, no. 11, pp. 3459–3468, Nov. 2005. 10.1109/TAP.2005.858842

[13] W-C Liu, "A design of a multiband CPW-fed monopole antenna using a particle swarm optimization approach," *IEEE Trans. Antennas Propagat.*, vol. 53, no. 10, pp. 3273–3279, Oct. 2005. 10.1109/TAP.2005.856339

[14] S. Cui and D. S. Weile, "Application of a parallel particle swarm optimization scheme to the design of electromagnetic absorbers," *IEEE Trans. Antennas Propagat.*, vol. 53, no. 11, pp. 3614–3624, Nov. 2005. 10.1109/TAP.2005.858866

[15] J. Perez and J. Basterrechea, "Particle swarm optimization and its application to antenna far-field pattern prediction from planner scanning," *Microwave Opt. Technol. Lett.*, vol. 44, pp. 389–403, Mar. 2005. 10.1002/mop.20648

[16] D. Boeringer and D. H. Werner, "Efficiency-constrained particle swarm optimization of a modified Bernstein polynomial for conformal array excitation amplitude synthesis," *IEEE Trans. Antennas Propagat.*, vol. 53, no. 8, pp. 2662–2673, Aug. 2005. 10.1109/TAP.2005.851783

[17] M. Donelli and A. Massa, "Computational approach based on a particle swarm optimizer for microwave imaging of two-dimensional dielectric scatterers," *IEEE Trans. Microwave Theor. Technol.*, vol. 35, pp. 1761–1776, May 2005. 10.1109/TMTT.2005.847068

[18] S. Ho, S. Yang, G. Ni, E. W. C. Lo, and H. C. Wong, "A particle swarm optimization-based method for multiobjective design optimizations," *IEEE Trans. Magn.*, vol. 41, no. 5, pp. 1756–1759, May 2005. 10.1109/TMAG.2005.846033

[19] W. Wang, Y. Lu, J. S. Fu, and Y. Z. Xiong, "Particle swarm optimization and finite-element based approach for microwave filter design," *IEEE Trans. Magn.*, vol. 41, no. 5, pp. 1800–1803, May 2005. 10.1109/TMAG.2005.846467

[20] M. Khodier and C. Christodoulou, "Linear array geometry synthesis with minimum sidelobe level and null control using particle swarm optimization," *IEEE Trans. Antennas Propagat.*, vol. 53, no. 8, pp. 2674–2679, Aug. 2005. 10.1109/TAP.2005.851762

[21] S. Genovesi and R. Mittra, "Particle swarm optimization for the design of frequency selective surfaces," *IEEE Antennas Wireless Propagat. Lett.*, vol. 5, pp. 277–279, 2006. 10.1109/LAWP.2006.875900

[22] N. Jin and Y. Rahmat-Samii, "Multi-objective particle swarm optimization for high performance array and reflector antennas," *IEEE Int. Symp. Antennas Propagat. Digest*, pp. 3293–3296, 2006.

[23] N. Jin and Y. Rahmat-Samii, "Real-number and binary multi-objective particle swarm optimization: a periodic antenna array design," *IEEE Int. Symp. Antennas Propagat. Digest*, pp. 3523–3526, 2006.

[24] Z. Bayraktar, P. L. Werner, and D. H. Werner, "The design of miniature three-element stochastic Yagi-Uda arrays in particle swarm optimization," *IEEE Antennas Wireless Propagat. Lett.*, vol. 5, pp. 22–26, 2006. 10.1109/LAWP.2005.863618

[25] R. Azaro, F. De Natale, M. Donelli, E. Zeni, and A. Massa, 'Synthesis of a prefracted dual-band monopolor antenna for GPS applications," *IEEE Antennas Wireless Propagat. Lett.*, vol. 5, pp. 361–364, 2006. 10.1109/LAWP.2006.880695

[26] M. Donelli, R. Azaro, F. De Natale, and A. Massa, "An innovative computational approach based on a particle swarm strategy for adaptive phased-arrays control," *IEEE Trans. Antennas Propagat.*, vol. 54, no. 3, pp. 888–898, Mar. 2006. 10.1109/TAP.2006.869912

[27] X. Shenheng and Y. Rahmat-Samii, "Boundary conditions in particle swarm optimization revisited," *IEEE Trans. Antennas Propagat.*, vol. 55, no. 3, pp. 760–765, Mar. 2007. 10.1109/TAP.2007.891562

[28] N. Jin and Y. Rahmat-Samii, "Advances in particle swarm optimization for antenna designs: real-number, binary, single-objective and multiobjective implementations," *IEEE Trans. Antennas Propagat.*, vol. 55, no. 3, pp. 556–567, Mar. 2007. 10.1109/TAP.2007.891552

[29] T. Huang and A. S. Mohan, "A microparticle swarm optimizer for the reconstruction of microwave images," *IEEE Trans. Antennas Propagat.*, vol. 55, no. 3, pp. 568–576, Mar. 2007. 10.1109/TAP.2007.891545

[30] K. C. Lee, C. W. Huang, and Y. H. Chen, "A microparticle swarm optimizer for the reconstruction of microwave images," *J. Electromagn. Waves Appl.*, vol. 21, no. 10, pp. 1353–1365, 2007. 10.1163/156939307783239474

[31] X. F. Liu, Y. B. Chen, Y. C. Jiao, and F. S. Zhang, "A microparticle swarm optimizer for the reconstruction of microwave images," *J. Electromagn. Waves Appl.*, vol. 21, no. 13, pp. 1819–1828, 2007.

[32] A. Semnani and M. Kamyab, "An enhanced method for inverse scattering problems using Fourier series expansion in conjunction with FDTD and PSO," *Progr. Electromagn. Res.*, vol. 76, pp. 45–64, 2007. 10.2528/PIER07061204

[33] W.-T. Li, X.-W. Shi, L. Xu, and Y.-Q. Hei, "Improved Ga and PSO culled hyhbrid algorithm for antenna array pattern synthesis," *Progr. Electromagn. Res.*, vol. 80, pp. 461–476, 2008. 10.2528/PIER07121503

[34] Z.-B. Lu, A. Zhang, and X.-Y. Hou, "Pattern synthesis of cylindrical conformal array by the modified particle swarm optimization algorithm," *Progr. Electromagn. Res.*, vol. 79, pp. 415–426, 2008. 10.2528/PIER07103004

[35] S. Selleri, M. Mussetta, P. Pirinoli, R. Zich, and L. Matekovits, "Differentiated Meta-PSO methods for array optimization," *IEEE Trans. Antennas Propagat.*, vol. 56, no. 1, pp. 67–75, Jan. 2008. 10.1109/TAP.2007.912942

[36] R. Poli, "Analysis of the publications on the applications of particle swarm optimization," *J. Artificial Evol. Appl.*, Article ID 685175, 10 pages, 2008. 10.1155/2008/685175

[37] Y. Rahmat-Samii and E. Michielssen, Eds., *Electromagnetic Optimization by Genetic Algorithms*, Wiley Series in Microwave and Optical Engineering, 1999.

[38] R. L. Haupt and S. E. Haupt, *Practical Genetic Algorithms*, Wiley-Interscience, 2004.

[39] R. L. Haupt and D. H. Werner, *Practical Genetic Algorithms*, Wiley-IEEE Press, 2007.

[40] J. Sun, B. Feng, and W. Xu, "Particle swarm optimization with particles having quantum behavior," *Proc. Cong. Evolutionary Computation*, CEC2004, vol. 1, pp. 325–331, June 2004.

[41] S. M. Mikki and A. A. Kishk, "Investigation of the quantum particle swarm optimization technique for electromagnetic applications," *IEEE Antennas Propagat. Soc. Int. Symp.*, pp. 45–48, vol. 2A, 3–8 July 2005. 10.1109/APS.2005.1551731

[42] S. M. Mikki and A. A. Kishk, "Quantum particle swarm optimization for electromagnetics," *IEEE Trans. Antennas Propagat.*, vol. 54, pp. 2764–2775, Oct. 2006. 10.1109/TAP.2006.882165

[43] J. M. Haile, *Molecular Dynamics Simulation*, New York, Wiley, 1992.

[44] W. Schommers, "Structures and dynamics of surfaces I," in *Topics in Current Physics*, vol. 41, W. Schommers and P. von Blanckenhagen, Eds., Springer-Verlag, Berlin, Heidelberg, 1986.

[45] M. Rieth, *Nano-Engineering in Science and Technology*, World Scientific, 2003.

[46] E. Kreyszig, *Advanced Engineering Mathematics*, 8th ed., Wiley, 1999.

[47] A. Papoulis, *Probability, Random Variables, and Stochastic Processes*, 3rd ed., McGraw-Hill, 1991.

[48] R. Penrose, *The Emperor's New Mind*, 1989.

[49] B. R. Gossick, *Hamilton's Principle and Physical Systems*, Academic Press, 1967.

[50] N. Iwasaki and K. Yasuda, "Adaptive particle swarm optimization using velocity feedback," *Int. J. Innovative Comput. Inform. Control*, vol. 1, no. 3, pp. 369–380, Sept. 2005.

[51] A. K. Bhattacharyya, *Phased Array Antennas*, Wiley-Interscience, 2006.

[52] A. Einestien, "On the movement of small particles suspended in a stationary liquid demanded by the molecular-kinetic theory of heat," *Ann. Phys.*, (Leipzig), 17, pp. 549–560, 1905. 10.1002/andp.19053220806

[53] J. Kennedy, "Probability and dynamics in the particle swarm," *Proc. Cong. Evolution. Computat.*, 2004, CEC2004, vol. 1, pp. 340–347, 19–23 June 2004. 10.1109/CEC.2004.1330877

[54] E. Nelsson, "Derivation of Schrödinger equation from Newtonian mechanics," *Phys. Rev.*, 150, pp. 1079–1085, 1966. 10.1103/PhysRev.150.1079

[55] L. Smolin, "Could quantum mechanics be approximation to another theory?" http://arxiv.org/abs/quant-ph/0609109, Sept. 2006.

[56] L. Smolin, *Life of the Cosmos*, Oxford University Press, 1999.

[57] S. Mikki and A. A. Kishk, "Improved particle swarm optimization technique using hard boundary conditions," *Microwave Optic. Technol. Lett.*, vol. 46, no. 5, pp. 422–426, Sept. 2005. 10.1002/mop.21004

[58] A. Carlisle and G. Dozier, "An off-the-shelf PSO," in *Proc. Workshop on Particle Swarm Optimization*, Indianapolis, IN, pp. 1–6, 6–7 April 2001.

[59] F. van den Bergh and A. P. Engelbrecht, "A cooperative approach to particle swarm optimization," *IEEE Trans. Evolution. Comput.*, vol. 8, no. 3, pp. 225–239, June 2004. 10.1109/TEVC.2004.826069

[60] A. Ratnaweera, S. K. Halgamuge, and H. C. Watson, "Self-organizing hierarchical particle swarm optimizer with time-varying acceleration coefficients," *IEEE Trans. Evolution. Comput.*, vol. 8, no. 3, pp. 240–255, June 2004. 10.1109/TEVC.2004.826071

[61] M. Wehr and G. Monich, "Detection of radiation leaks by spherically scanned field data," in *Proc. 10th Int. Zurich Symp. Technol. Exhb. EMC*, pp. 337–342, 1993.

[62] M. Wehr, A. Podubrin, and G. Monich, "Automated search for models by evolution strategy to characterize radiators," in *Proc. 11th Int. Zurich Symp. Technol. Exhb. EMC*, pp. 27–34, 1995.

[63] J. Rammon, M. Ribo, J.-M. Garrell, and A. Martin, "A genetic algorithm method for source identification and far-field radiated emissions predicted from near-field measurements for PCB characterization," *IEEE Trans. Electromagn. Comp.*, vol. 43, no. 4, pp. 520–530, Nov. 2001.

[64] B. M. Kolundzija, J. S. Ognjanovic, and T. K. Sarkar, *WIPL-D: Electromagnetic Modeling of Composite Metallic and Dielectric Structures, Software and User's Manual,* Reading, MA: Artech House, 2000.

[65] T. S. Sijher and A. A. Kishk, "Antenna modeling by infinitesimal dipoles using genetic algorithms," *Progr. Electromagn. Res.*, vol. 52, pp. 225–254, 2005. 10.2528/PIER04081801

[66] E. Collin, *Foundation for Microwave Engineering*, McGraw-Hill, 1966.

[67] D. Kajfez, *Q Factor*, Vector Fields, 1994.

[68] S. M. Mikki and A. A. Kishk, "Physical theory for particle swarm optimization," *Progr. Electromagn. Wave Res.*, vol. 75, pp. 171–207, 2007. 10.2528/PIER07051502

[69] S. M. Mikki and A. A. Kishk, "Hybrid periodic boundary condition for particle swarm optimization," *IEEE Trans. Antennas Propagat.*, vol. 55, no. 11, Nov. 2007. 10.1109/TAP.2007.908810

[70] T. Huang and A. Sanagavarapu, "A hybrid boundary condition for robust particle swarm optimization," *IEEE Antennas Wireless Propagat. Lett.*, vol. 4, pp. 112–117, 2005. 10.1109/LAWP.2005.846166

[71] P. A. Lindsay, *Quantum Mechanics for Electrical Engineers*, McGraw-Hill, 1967.

Index

Printed in the United States
by Baker & Taylor Publisher Services